AQUACULTURE SOURCEBOOK

A Guide to North American Species

AQUACULTURE SOURCEBOOK

A Guide to North American Species

Edwin S. Iversen
and
Kay K. Hale

An AVI Book
Published by Van Nostrand Reinhold
New York

An AVI Book
(AVI is an imprint of Van Nostrand Reinhold)

Library of Congress Catalog Card Number 92-12536

ISBN 0-442-00992-5

Manufactured in the United States of America

Published by Van Nostrand Reinhold
115 Fifth Avenue
New York, New York 10003

Chapman and Hall
2-6 Boundary Row
London, SE 1 8HN, England

Thomas Nelson Australia
102 Dodds Street
South Melbourne 3205
Victoria, Australia

Nelson Canada
1120 Birchmont Road
Scarborough, Ontario M1K 5G4, Canada

16 15 14 13 12 11 10 9 8 7 6 5 4 3 2 1

Library of Congress Cataloging-in-Publication Data

Iversen, Edwin S.
Aquaculture sourcebook : a guide to North American species / by Edwin S. Iversen and Kay K. Hale.
p. cm.
"An AVI book."
Includes bibliographical references (p.) and index.
ISBN 0-442-00992-5
1. Aquaculture—North America—Handbooks, manuals, etc. I. Hale, Kay K. II. Title.

SH33.I94 1992
639.8´097—dc20

92–12536
CIP

Contents

Preface

Since the late 1970s, interest in aquaculture—the fresh-, brackish-, and salt-water farming of fish and shellfish—has grown by leaps and bounds. There has been a simultaneous boom in the literature of aquaculture.

Both authors receive frequent requests for information on the aquacultural potential of organisms. Some are from people with little or no background in fishery science; they want a brief, easy-to-scan text outlining information on species they might successfully produce in a gertain geographic area or under special conditions. Others are from individuals with more scientific knowledge and/or preliminary market research; they are seeking further details on a particular candidate species.

This book is designed to satisfy requests of both kinds, as well as to serve as a text for introductory courses in aquaculture. It is a ready reference to information on aquaculture, using terminology that can be understood by the layperson as well as by the student and fishery scientist. Over a hundred species are discussed in an easy, standardized format. Each of these species entries covers the organism's scientific and common names, description and distinctive characteristics, habitat range, reproduction and development, age and growth, food and feeding, parasites and disease, predators and competitors, aquacultural potential, and regions where each of these species are (or have been) farmed and/or researched. The most important biological, environmental, and other pertinent requirements necessary for successfully raising each species are emphasized; possible pitfalls and disadvantages presented; and the most useful, available current literature listed, providing broader coverage for those readers who may need it.

The design of aquacultural facilities and financial aspects are not covered in any detail; neither are the legal contraints involved in starting up an aquacultural facility, nor the processing, marketing, or transportation of the product. These aspects, though certainly important, are beyond the scope of this work: Current publications on these additional subjects are listed in the Selected References near the end of the book.

The majority of illustrations of fish species are from detailed drawings by H. L. Todd, and drawings of the invertebrate species are by J. H. Emerton and associates; these appeared in G. B. Goode, *The fisheries and fishery industries of the United States,* published during 1884–7 by the U.S. Government Printing Office. The drawings of the rotifer, copepod, and sea hare are reprinted by permission of Wadsworth, Inc., from R. A. Pimental (1967), *Invertebrate identification manual,* New York, Reinhold.

Additional credit for photographs and drawings goes to the following:

International Pacific Halibut Commission (early development of halibut);
A. P. Andriashev (Arctic char);
C. N. D'Asaro (lugworm);
I. Flye (bloodworm);
P. S. Galtsoff (mangrove oyster);
J. Z. Iversen (brine shrimp, pen shell, tubifex worm, and the life-cycle drawings of penaeid shrimp, milkfish, and mullet);
J. A. Oliver (alligator, crocodile);
J. Randall (queen conch);
S. E. Smith (hypothetical bony fish);
J. F. Storr (sponges); and
A. Williams (West Indian king crab).

Our thanks are also extended to the various state, federal, and international organizations whose publications have provided illustrations used in this volume. We thank Don Heuer of the Rosenstiel School of Marine and Atmospheric Science (RSMAS) Printing and Photo Services for his expert help with the illustrations, and Assistant Librarian Helen D. Albertson for her efficient and good-natured interlibrary loans. We also gratefully acknowledge the anonymous reviewer whose suggestions for reorganization were incorporated into the final draft manuscript.

We would like to express out appreciation to the following editorial staff members of Van Nostrand Reinhold for their cooperative and conscientious assistance in the publication of this work: Dr. Eleanor Riemer, Michael Beck, and Michael Gnat.

1

Introduction

Despite documentation that various types of aquaculture were practiced as early as 2000 B.C. in the Far East, interest in the Western world developed rapidly only about thirty years ago. Optimistic articles in the news media predicted how this "newfound" protein source could literally feed the world. Based on these early reports, Western entrepreneurs had great expectations for profitable ventures. At a time when the buzz words in many developed countries were "new ventures" and "diversification," large corporations allotted substantial portions of their financial resources to aquaculture. Unfortunately, at this time the biotechnology of aquatic organisms and aquacultural field trials for many species were unavailable, and there was little significant progress. Accounts of failed ventures were commonplace, and examples of Western aquacultural success were few.

The types of aquaculture successful in the Far East were, in many cases, small subsistence and part-time operations conducted in order to feed members of the farmer's immediate family; there was little or no profit motive. Western entrepreneurs faced the grim conclusion that most aquaculture ventures were big money losers. They had not recognized the high level of economic risk, and many were unaware of how extremely limited biotechnological knowledge was for many species. Success required information.

After years of experience, we now see new optimistic, yet more realistic, outlooks. Regarding private aquacultural ventures, Dr. P. Larkin—a noted fishery biologist with the Institute of Animal Resource Ecology, University of British Columbia—predicts that "over the next 50 years, aquacultural production will equal, if not surpass, wild production of fish" (Larkin 1988). For this prediction to be realized, present world fresh- and saltwater aquacultural production—about 13 million metric tons (mmt)—must increase to about 75 mmt (assuming world capture-fishery production remains constant). This means that world aquacultural production would have to increase about 1.2 mmt per year.

Dr. H. Rosenthal (1985) predicts that world aquacultural production will increase to about 30 mmt by the year 2000. Given the U.N. Food and Agricultural Organization's (FAO) 1985 world-production estimate of 13 mmt, Rosenthal's prediction requires an increase of 17 mmt in fifteen years, or about 1.13 mmt a year. This figure, close to that of Larkin, also reflects an optimistic future for aquaculture. FAO scientists (quoted in Shrimp farm boom! 1988) estimate that, by the year 2000, annual aquacultural production of freshwater prawns and tropical marine shrimp alone will exceed 1 mmt. The many reasons for increased maricultural production are reviewed by Ryther (1981).

It should be noted that estimating worldwide aquacultural production is problematic for several reasons: the enormity of the area estimated, the scarcity of data, the diversity of aquacultural and data-reporting practices, and so on. Still, there are good reasons to accept the optimism reflected in these predictions, not the least of which is that they are made by highly qualified fishery scientists with years of experience. Moreover, present-day aquaculture has clearly benefited from numerous biotechnological advances and accumulated information.

SOURCES OF AQUACULTURAL INFORMATION

Until recently, there were very few reliable sources of information on aquaculture available. The number of aquaculture books and monographs has doubled every five years since 1961 (Maclean 1988); in addition, there are now numerous journals dedicated to aquaculture, as well as many others in the fields of agriculture, economics, and medicine that publish aquaculture articles.

Requests for aquacultural information from the U.S. Department of Agriculture's (USDA) National Agriculture Library have increased sharply, from 250 in 1985 to 4,200 requests in 1989. A number of government and private institutions also publish "how-to" handbooks and manuals or offer courses in aquaculture. These courses, which are advertised in aquaculture journals and magazines, range from college level to those requiring little or no formal education. Some even include hands-on rearing of shrimp and finfish.

The new literature provides invaluable background on aquacultural biotechnology; consequently, more realistic articles are now available on the practical problems and constraints of starting and operating aquacultural ventures. Rosenthal (1985) identifies environmental degradation, disease control in aquatic species, human health, and competition for resources as important impediments to successful aquaculture. He also notes that research gaps in science and technology still exist. Shupe (1982) describes competing interest groups (salmon fishermen vs. salmon ranchers and farmers) as well as legal constraints (regulatory complexities of private vs. public rights).

Due to a variety of social, legal, and economic constraints, technological advances made in U.S. and Canadian aquaculture frequently cannot be applied to Third World countries; but such advances have provided impetus to aquacultural development where regulations are less stringent (Ryther 1984).

ABOUT THIS BOOK

This volume emphasizes aquaculture species raised in North America. General discussions of the status of aquaculture in the United States, Canada, and Mexico are provided in Chapter 2. Subsequent chapters provide useful information on individual aquacultural food (Chapter 3) and nonfood (Chapter 4) species: their scientific and common names, description and distinctive characteristics, habitat range, reproduction and development, age and growth, food and feeding, parasites and disease, predators and competitors, aquacultural potential, and regions where each of these species are (or have been) farmed and/or researched. For each species discussed, we also list reference works helpful in evaluating the likelihood of successful farming; most if not all of these are current and easily available. The sequence in which species are presented in these chapters runs parallel to the group classification outlined in Appendix B; finfish species are alphabetized by keyword (largemouth *bass, buffalo* fish, Arctic *char,* red *drum,* etc.).

Scientific and Common Names

The scientific and common names featured in this book are those recommended by the scientific community. Additional common names may be listed together with that recommended for a species (or group of species) that is of considerable economic importance. Note that information presented in a discussion of an individual species pertains to the *first-listed species only:* Although the life history of related species listed will resemble that of the first-listed species, their biological attributes, growth habitat, and so on may well vary.

Where several closely related species have the same general biological attributes, geographic range, and potential for aquaculture, their scientific and common names are listed, and their attributes discussed, in *group summaries*. We then select one of the more important of these related species and investigate its desirability for commercial aquaculture. For example, several species of warmwater (penaeid) shrimp are highly successful in aquaculture: *Penaeus vannamei, P. monodon,* and *P. japonicus*. Although the latter two are not native to the Western Hemisphere, we include material on their life histories because they are farmed on a large scale and have been transplanted widely. Additional species of *Penaeus* shrimp, some of which are listed in the summary, have rather similar life histories; however, variations in their biology and ecology must be taken into account.

Description and Distinctive Characteristics

Information on size, shape, coloration, and distinguishing characteristics is provided together with an illustration of the species or species group. (In a few designated instances, the illustration may be of a closely related species.) These descriptions and illustrations are intended only to give the reader a general idea of the species' appearance, not to enable precise scientific identification.

Habitat Range

For each species highlighted, the generally accepted habitat range is given. We also list areas where the species has been transplanted; only the larger regions are listed for species that have been transplanted widely. A reliable source for this information will be cited if available.

It is important to remember that species have an optimal temperature range for good growth, as well as maximum and minimum temperatures they can tolerate. In addition, some species require a certain salinity range in order to thrive. To transplant a species outside of its normal ranges is impractical unless unusual conditions exist: For instance, warm-water fish can be raised in Idaho during the winter only because of its geothermal springs. Normally, however, species exist in small numbers at the extremes of their geographic ranges because environmental conditions there are frequently marginal for the growth and survival for the species. When even one essential environmental condition reaches a limiting level—preventing a species from flourishing, but still allowing it to survive—even a minor deterioration in this condition may promote the death of the species. Conversely, a slight improvement in this environmental condition is likely to be followed by an upsurge in its well-being. It is the tipping of this delicate environmental balance that often leads to success or failure in aquacultural ventures. This explains the observed geographic distribution of the species: Where conditions are just right, the species thrives; but as its distribution extends in any direction from this optimal center, one or more environmental conditions become limiting, and the species cannot survive. This is why great care is required when choosing a candidate species and locating an aquacultural facility: The location should provide the species with living conditions as close as possible to optimum.

Table 1.1 gives examples of aquaculture species that undergo short- or long-term habitat changes (salinity, water to land) during their natural life cycles. We do not include species that are *stenohaline* (i.e., that live in a narrow range of salinity), nor those that strictly require fresh water. Within some groups certain individual species may have a greater tolerance to salinity than do others. Also, the ability to survive and grow well in either low or high salinity may be a function of age of the animal; for example, warm-water (penaeid) shrimp usually grow better in low salinity when young and in high salinity when adults.

Table 1.1. Examples of Species That Undergo Short- or Long-Term Habitat Changes during Their Life Cycles or That Can Tolerate Environmental Changes[1]

ALGAE

Microalgae (nori)

INVERTEBRATES

Brine shrimp
Copepods
Crabs (blue crab, giant land crab, land hermit crab)
Oysters (mangrove oyster)
Rotifers
Shrimp/prawns (brown shrimp, giant tiger shrimp, kuruma prawn, pink shrimp, white leg shrimp)
Snails
Tubifex worms

VERTEBRATES

FINFISHES

Arctic char
Bass (striped bass)
Catfish (walking catfish)
Eels (*Anquillia* spp.)
Florida pompano
Milkfish
Mullet (striped mullet)
Redfish (red drum)
Salmon (Atlantic salmon, Pacific salmon [king salmon, coho salmon])
Shad (American shad, threadfin shad)
Snook
Sturgeon
Tilapia
Trout (rainbow [steelhead] trout, seatrout)

REPTILES

Alligators
Crocodiles
Turtles (terrapins)

AMPHIBIANS

Frogs

[1]Primarily changes in salinity, though some species can tolerate a moderate temperature range.

Marine animals can be raised far inland, where natural seawater is unavailable, by mixing commercially available seawater salts with fresh water to produce an acceptable substitute. This process is suitable for small-scale aquaculture, such as rearing ornamental fish, but is far too expensive for commercial aquaculture of food fish.

Some species, such as salmon, are classified as *anadromous* because they enter fresh water to spawn. Because of this tolerance to a wide range in salinity, anadromous species will grow and reproduce in fresh water (e.g., landlocked salmon in freshwater Alaskan lakes, salmon transplanted to the Great Lakes). Species that migrate to sea from fresh water to spawn (e.g., eels) are called *catadromous*. Some biologists use the term *quasi-catadromous* for species that spawn in salt water but whose very early life stages enter estuaries to find conditions suitable for rapid growth (e.g., blue crab, menhaden, mullet warm-water shrimp, red drum). Some species (e.g., turtles and alligators) leave the water to lay their eggs on land; others (land hermit crabs, giant land crabs) leave the land to lay their eggs in water. Certain aquatic species (e.g., walking catfish) even travel over land in search of better habitats.

With so many possible variations in habitational requirements, the importance of investigating thoroughly those of any candidate aquaculture species cannot be overstressed.

In recent years it has become abundantly clear that the haphazard introduction of plants and animal species new to some area, under the assumption that they would produce a species to the benefit of humankind, can lead to nightmarish situations. Competition for food and space with local species, introduction of diseases, and the disruption of predator–prey relationships are just some of the problems that have resulted. Today, a concerted effort is being made by regulatory agencies to prevent further harmful transplantations. Agency guidelines recommend the denial of such introductions unless there is unambiguous evidence that no harm to the local ecology will ensue. (In the case of species to be held in enclosures, no harm must ensue should they escape confinement.) Even then, the individual or organization receiving the exotic species—usually a state or federal research laboratory—must adhere to very strict conditions. In general, transplantation of exotic species requires considerable study. It should be investigated thoroughly and then carried out *only with proper authorization* (governmental permits, etc.).

Reproduction and Development

There are three questions that an aquaculturist must be able to answer with regard to most candidate species:

1. Will the species breed under artificial conditions?
2. How fecund is the species (i.e., how many eggs do they lay)?
3. How long do the early developmental stages last (i.e., what is the larval life)?

Aquaculturists are said to have "control over reproduction" when they can induce species to *breed under artificial conditions*. (The life cycle of the species is then described as "closed.") Under these circumstances, a dependable supply of young, all from known parents, may be produced by brood stock. Otherwise, young stages will have to be obtained from nature; this has many drawbacks, including undependable supplies of young (which, in turn, depend on natural conditions), numerous parasites, introduction of predators and other undesirable animals difficult to cull out, and usually a high cost of collection.

Fecundity becomes problematic at both ends of the spectrum: Species that lay relatively few eggs require more brood fish to produce sufficient young for stocking. However, fish that are highly fecund generally have low survival rates during their early stages; hence, again, many brood fish must be maintained at the aquaculture facility.

An *extended larval life,* such as that demonstrated by the spiny lobster (9–12 months), is a severe disadvantage to their commercial success in aquaculture. Holding these tiny, delicate larval stages and providing the correct feed and environmental conditions over long periods can be very costly as well as risky.

Having a consistent, year-round supply of young seed stock of fish or shellfish means inducing adult individuals to mature and mate in captivity. This is usually done by placing them in a *maturation facility,* where light intensity, photoperiod, water temperature, feed, and so on are controlled to provide conditions as similar as possible to those under which they spawn naturally.

Once mature (or nearly mature) fish or shellfish have been manipulated to spawn, the resulting sperm and ova are held under artificial conditions that encourage fertilization, hatching, and high survival of the larvae. Such *hatchery* operations imply early release of the *fry* (young juveniles) into the wild to fend for themselves, with the anticipation that many will be subsequently caught by fishermen. Unfortunately, as with the young of wild parents, only a tiny fraction of the released fish survive to maturity; the rest either (1) are consumed by the millions by a variety of predators, (2) starve to death, or (3) are carried by currents into unfavorable environmental conditions.

Therefore, in addition to maturation facilities and hatcheries, high-intensity aquaculture (intensive fish farming) usually requires that fish be held in enclosures, or *grow-out facilities,* for much longer periods of time. There they are provided a high degree of husbandry, including protection from predation and disease and suitable food, until they are ready to be harvested for market.

Age and Growth

Estimates of age and growth are critical to predicting the success of commercial aquaculture ventures. Since long-lived, slow-growing animals/plants are costly to hold, and are subject to mortality over a longer period, they frequently are unprofitable to raise.

The *growth rate* is the time required for an animal to reach a certain life stage or size (usually market size). Some researchers include length of larval life when determining the growth rate; others do not. It may be based on wild specimens or those reared in captivity, and may differ considerably for fish or shellfish well fed and provided an optimal grow-out environment in an aquaculture or research facility versus those living in the wild. For instance, marine lobsters in New England require five to eight years to reach a marketable size (1 lb) in the wild; however, the same size can be reached in about two years when conditions—especially temperature—are controlled. Because growth rates may be calculated based on specimens held under a variety of conditions, they should only be used as estimates of the times required to reach a life stage or size.

The following environmental factors interact with culture conditions to affect the growth and survival of plants and animals reared in controlled conditions:

geographic area	water temperature
pond density	kinds and quality of feeds used
feeding rate	rate of water exchange
sizes and mixture of sizes	

However, even if conditions in an artificial facility are optimalized for a given species, there will still be limits on the growth rates, attainable size, fecundity, and so on: By providing what seem to be ideal conditions, one may expect only to reach, *not exceed,* these limitations. However, the attributes of species induced to spawn and be raised in captivity may be enhanced by *artificial selection;* this has been used widely for domestic land animals and, in a few cases, for finfishes (e.g., salmon and ornamental fish [goldfish]).

Food and Feeding

A few definitions are in order here, to avoid confusion. A species' *food* is considered to be the materials eaten by it under *natural* conditions. *Feeds* are usually considered to be materials *manufactured or prepared,* and are available in a number of forms (wet, dry, encapsulated, flaked, pelleted, extruded, etc.). Dry prepared feeds normally do not require refrigeration for short-term storage. The term *practical feeds* has been suggested to include a number of feeds produced by commercial processes, but we prefer the term *prepared feeds,* which is more accurately descriptive; moreover, in many small-scale operations feeds are prepared on the grounds of the facility, not commercially manufactured. *Constituted* feeds comprise certain components in specific proportions, both in terms of ingredients and nutritional analysis. Fish meal, grain meal, meat meal, and additives are compounded to make prepared feeds. Proteins (plant or animal) usually range between 20 and 35 percent, with fats, carbohydrates, minerals, medicines, vitamins, and other minor constituents making up the rest.

Satisfying the nutritional requirements of an aquaculture species is a major consideration in determining profits in most ventures. It is much cheaper to raise bivalve mollusks (e.g., oysters, mussels, scallops), which feed low on the food web (chain) by filtering plankton in nearshore waters where favorable current systems bring ample food to them, than it is to raise carnivores, which are held in enclosures and require meat protein.

Because the cost of feed is one of the big-budget items, *feed conversions* are calculated; these are ratios indicating how many pounds of feed are needed to produce a pound of the cultured species. (A conversion of 2:1 means that 2 lb of feed are required to produce 1 lb of fish or shellfish.) As a rule, an actively swimming fish like the dolphin fish requires much more feed to produce a pound of fish flesh than does a less active fish like the channel catfish. This is because a substantial portion of the feed intake of active fish goes into maintenance (swimming). Likewise, female tilapia that spawn frequently have poorer feed conversions than do males because of the energy that goes into egg production. Obviously, fish that have poor feed conversions and require high-priced feeds reduce the profit margin in commercial production.

Prepared feeds and natural foods may be used together in aquaculture facilities. Due to the high density of fish in enclosures, the prepared feeds may serve as *supplemental feeds* if the carrying capacity of the pond is inadequate to satisfy the nutritional requirements of the species raised. For instance, studies with golden shiners have shown that supplemental feeding increased production by about 50 percent over ponds that were only fertilized. Supplemental feeds can also provide constituents that may be low or absent in the natural diets but that are required for the species raised during some period of its life: Many fish and shellfish change their diet as they pass through their various life stages (larvae, juveniles, adults, spawners).

In addition to fulfilling nutritional requirements, it is important to provide feeds of proper size and appetizing appearance, and in the correct amounts. Depending on the feeding habits of the fish reared, either floating or sinking pellets can be used; these are available in a wide range of sizes. To keep them together in water, pelleted feeds are made with binders, which the aquaculture species must be capable of digesting. For species "new to aquaculture," prepared feeds frequently are not available: New feeds are costly to manufacture, and the feed companies must be ensured a market sufficient to repay the cost of the experimentation and development entailed.

In some facilities, such as hatcheries, nurseries, and grow-out ponds, the aquaculture species may either be raised on prepared feeds, such as pellets, or on foods *cultured* to feed it. The disadvantages and advantages of each system are discussed in some of the recent handbooks (e.g., McLarney 1984). In some situations the only food available that will ensure good growth and survival of the aquaculture species is live food; however, the cost and extra effort that go into rearing and maintaining cultures of live forage organisms may be considerable. In this book we have included only a few examples of species cultured as food for aquaculture species: microalgae, copepods, daphnia (water fleas), and brine shrimp. There are many more presently being used or that have potential for use in aquaculture, especially in fresh water. (See Chapter 4 on nonfood species.)

Parasites and Disease

Fish in captivity are stressed by high density, suboptimal temperatures, and, occasionally, low dissolved oxygen, causing many to become susceptible to disease. Microorganisms (virus, bacteria, fungi) are responsible for many diseases, as is improper nutrition. Some kinds of fungus cause disease directly in fish; others are secondarily invaders in lesions, injuries, and so on. Poor nutrition by itself can cause serious diseases; in combination with stress caused by parasites and adverse environmental conditions, it can destroy a fish crop.

For species that have not been raised in artificial conditions, the parasites and diseases recorded in the literature are frequently those from wild fish and shellfish. The reader may be shocked by the numbers and different types of parasites that either *infect* (internal) or *infest* (external) many wild fish throughout their range (see Appendix G). Fortunately, many of these parasites are excluded from aquacultural facilities because potential host fish do not eat the infected prey fish or shellfish that are normally intermediate hosts in the wild. Constituted (pelleted) diets prevent heavy infections of macroparasites, such as flatworms and roundworms.

Predators and Competitors

Some species are heavily preyed upon by other animals (such as those shown in Appendix F). Attempts to exclude predators may be costly and difficult; some fish-eating birds that can cause serious losses to fish farmers are even protected by U.S. law. Protective measures, such as making loud noises, displaying artificial enemies of the predatory birds (e.g., imitation owls), and even covering pond areas with webbing may not significantly reduce this predation.

Aquacultural species that are cannibalistic (e.g., many crustaceans, especially crabs) must be sorted to size and the different sizes separated to reduce cannibalism.

Some species are affected by competition for the necessities of life (food, dissolved oxygen, etc.), and if competition is keen, growth will be reduced and mortality increased.

Aquacultural Potential

This is the most important section for individuals interested in setting up an aquacultural facility using one or more of the species described in this book. In a nutshell, a species is a good candidate for aquaculture if it meets the following criteria:

1. The species has a strong market demand, a favorable selling price, and a common name that is well known to the consumer.
2. The biology of the species is well known and there is a considerable body of literature on all its aspects.
3. The species has been successfully farmed *commercially,* that is, not just in research facilities but also in for-profit operations.

4. The species lends itself to confinement and handling without harm.
5. Reproduction has been carried out in captivity (i.e., there is control over reproduction, the life cycle is closed [see "Reproduction and Development" above]).
6. The larval life is short.
7. Growth is rapid, and survival to a market size is high.
8. The species feeds low on the food chain and/or has a very favorable conversion ratio of feed to fish flesh.
9. Few parasites or diseases are known to cause mass mortality in the culture of the species; for those that do exist—and there are some for most species—there are known methods of preventing and controlling them.
10. Predators, cannibalism, and inter- and intraspecific competition do not seriously affect mass rearing of the species.

In determining the suitability to aquaculture of a given species, it is important to consider the characteristics not only of that species but also of the larger group to which it belongs. For example, crustaceans (shrimp, crabs, and lobsters) tend to be more difficult to raise than mollusks or finfish because they have long larval life stages and are very cannibalistic; on the other hand, they frequently have very strong market demand.

Another point to keep in mind is that some species of fish and shellfish can be raised for *more than one market.* Tilapia, for instance, is marketable as human food but can also be sold at small sizes for fish bait. Moreover, some species in this group are very colorful and are marketed as ornamental fishes. Many states require a special permit for the possession and sale of tilapia species, which are exotics; in other states, such possession and/or sale is illegal.

In the "Aquacultural Potential" sections, we try to emphasize difficulties, as they *presently exist,* that an individual might encounter when trying to culture the subject species for economic gain. We hope this cautionary information will forestall financial losses such as those that have occurred in numerous aquacultural ventures over the past several decades in many developed countries, including the United States. However, our evaluations of various species and species groups are, at best, only a general guide. The potential of each candidate species can change dramatically as new biological information and technology develops. Levels of seafood production—from local capture fisheries as well as imports from foreign capture and culture fisheries—have an impact on the aquacultural potential of some species, as do overall market economic fluctuations.

Competition between culture and capture fisheries, real or imagined, is rather long-standing. However, although there used to be relatively little competition among culture fisheries worldwide, nowadays the production of farmed species from several different countries may compete for a share of the world market.

For example, Norway, Scotland, Chile, and Canada actively compete in the salmon market, and Asian and South American countries actively compete in the shrimp market. The obvious result of this increased production is a reduced margin between production costs and market prices received—that is, reduced profit. Management must be efficient and production costs low in order for these businesses to survive.

Regions Where Farmed and/or Researched

We have tried to locate some of the more important areas where information useful to aquaculturists might be obtained. Hence, for each species presented, locations are given for where aquacultural research is being (or has been) conducted, as well as for where governmental or commercial fish farms are (or were) in operation. Information regarding new ventures or the termination of old ones is difficult to come by, and it may not always be clear why a facility is no longer engaged in aquacultural research or production. However, the fact that a species is either actively being considered for aquaculture or is already successfully being raised in a given geographic area may provide some guidance as to the type of location and climate suitable for that species.

Although this book emphasizes aquacultural products and practices of the United States, Mexico, and Canada, species native to and successfully raised in other regions are also included, as are those that have been widely transplanted. Information regarding species farmed or researched outside of North America may have some bearing on the potential of a closely related indigenous species.

ASPECTS OF GOOD AQUACULTURAL PLANNING

We have already stressed the importance of rearing suitable animals in a suitable habitat. Many early aquaculture ventures failed to consider habitat suitability in their planning, and were uninformed or careless when selecting species to culture. Operations were begun with little biological knowledge and few if any successful commercial aquacultural trials on record. Many aquaculturists either did not understand or inadequately anticipated losses caused by biological problems. Today, these problems can be greatly reduced for many (though not all) species, thanks to years of biological research and aquacultural trials.

Aquaculture is categorized as a high-risk venture. The majority of large aquacultural facilities require outdoor grow-out ponds, with the stock at the mercy of the weather. Unusually bad weather can wipe out entire ponds of fish that have been raised at great care and expense. Careless workers and lack of attention to details can duplicate the ravages of weather. Many early U.S. trials were doomed by the high profit required on investments, together with high labor, land, and feed costs. However, thanks to recent U.S. federal, state, and private research programs, there is now a tremendous store of valuable biological and practical data that can be applied when planning aquaculture ventures, whether within or outside the United States.

First read as much as you can find on the aquaculture of the species in which you are interested; them talk with as many informed individuals as possible, and contact associations (see Appendix H) that can provide you with additional information. Be sure to obtain a clear idea of the size of the market and the extent of market competition from capture fisheries. Visits to extant commercial aquacultural facilities can be quite helpful; however, many such businesses are naturally reluctant to invite potential competitors to examine their operations.

Once a suitable aquacultural species and facility location have been selected, carefully and thoroughly investigate the legal and business aspects of the proposed venture. Prior to any substantial financial investment in an operation, it is vital to research thoroughly the federal, state, and local laws that pertain to the design, establishment, and operation of an aquacultural facility. Several state and federal agencies can be of assistance here, including departments of natural resources, state and national Sea Grant programs, and extension services. Many recent aquacultural ventures have been impeded by insufficient research into legalities.

Adequate capitalization is also very important. Be sure to take into account the time required to obtain permits and set the operation in motion. Note that once startup has been achieved, *several years* of operation may be required in which to work out all the problems; during this phase there are many expenses yet little or no income. Economic analyses are a requisite guide for industry investment decisions.

There is no substitute for a well-designed business plan to help ensure success. Today, computers have made the financial planning of an aquacultural facility easy and practical. All the necessary number-crunching—testing the financial viability of various facility sizes, calculating startup and operating costs, determining break-even points, and so forth—can now be done precisely and in short order. This is not to say that common sense and caution should be thrown to the wind; but given good input data, this kind of planning is quite valuable and can help one avoid business failure.

Despite the numerous difficulties in planning and operating commercial aquacultural ventures, there are today a great many successful ones, and the future of aquaculture looks promising. It has been suggested that recognizing aquaculture *as a form of agriculture*—with which it shares the threat of losses due to weather, disease, and crop-feeding animals—would help resolve the industry's biggest remaining practical problems (e.g., controlling legally protected predators, obtaining disaster loans) and enhance the chances for greater development.

REFERENCES

Allen, L. J., and E. C. Kinney, eds. 1981. *Proceedings of the Bio-Engineering Symposium for Fish Culture.* Bethesda, MD: Fish Culture Section of the American Fisheries Society.

Avault, J. W. 1988. "New" species for aquaculture. *Aquaculture Magazine* 14(2):53–5.

Balchen, J. G., ed. 1987. *Automation and data processing in aquaculture: Proceedings of the IFAC symposium, Trondheim, Norway, 1986.* Oxford: Pergamon Press.

Courtenay, W. R., and J. R. Stauffer. 1984. *Distribution, biology, and management of exotic fishes.* Baltimore: Johns Hopkins University Press.

Fuss, J. T. 1983. Using single board computers for control applications. *Fisheries* 8(5):8–13.

Hatch, U., S. Sindelar, D. Rouse, and H. Perez. 1987. Demonstrating the use of risk programming for aquaculture farm management: The case of penaeid shrimp in Panama. *Journal of the World Aquaculture Society* 18(4):260–9.

Larkin, Peter A. 1988. The future of fisheries management: Managing the fisherman. *Fisheries* 13(1):3–9.

Lewis, R. 1988. Fish: New focus for biotechnology. *BioScience* 38(4):225–7.

Ling, S. W., and T. J. Costello. 1979. The culture of freshwater prawns: A review. In *Advances in aquaculture: Papers presented at the FAO Technical Conference on Aquaculture, 1976,* ed. T. V. R. Pillay and W.A. Dill, pp. 299–305. Farnham (U.K.): Fishing News Books.

Maclean, J. L. 1988. The growth of fisheries literature. *Naga, the ICLARM Quarterly* 11(1): 3–4.

McLarney, W. O. 1984. *The freshwater aquaculture book: A handbook for small scale fish culture in North America.* Point Roberts, WA: Hartley & Marks.

Rogers, G. L., T. L. Richard, S. L. Klemetson, and A. W. Fast. 1986. Application of spreadsheets in aquaculture. *Aquaculture Magazine* 12(2):51–4.

Rosenthal, H. 1985. Constraints and perspectives in aquaculture development. *GeoJournal* 10(3):305–24.

Ryther, J. H. 1981. Mariculture, ocean ranching, and other culture-based fisheries. *BioScience* 31(3):223–30.

Ryther, J. H. 1984. Biotechnology: A potential U. S. contribution to mariculture. In *Biotechnology in the marine sciences: Proceedings of the First Annual MIT Sea Grant Lecture and Seminar,* ed. R. R. Colwell, A. J. Sinskey, and E. R. Pariser, pp. 123–32. New York: John Wiley.

Shrimp farm boom! 1988. *Fish farming international* 15(3):1.

Shupe, S. J. 1982. *Coastal aquaculture: Protein, profits and problems for a hungry world.* Corvallis, OR: Sea Grant College Program, Oregon State University.

Sindermann, C. J. 1986. Strategies for reducing risks from introductions of aquatic organisms: A marine perspective. *Fisheries* 11(2):10–15.

Sindermann, C. J., and D. V. Lightner, eds. 1988. *Disease diagnosis and control in North-American marine aquaculture,* 2nd rev. ed. Amsterdam: Elsevier.

Tiddens, A. 1990. *Aquaculture in America: The role of science, government and the entrepreneur.* Boulder, CO: Westview Press.

Webber, H. H. 1972. The design of an aquaculture enterprise. *Proceedings of the Gulf and Caribbean Fisheries Institute* 24:117–25.

Webber, H. H. 1973. Risks to the aquaculture enterprise. *Aquaculture* 2(2):157–72.

Webber, H. H., and P. F. Riordan 1976. Criteria for candidate species for aquaculture. *Aquaculture* 7(2):107–23.

2

U.S., Canadian, and Mexican Aquaculture

U.S. AQUACULTURE

The beginning of fish culture in the United States is somewhat clouded by the passage of time. A Rev. John Bachman claimed to have hatched freshwater trout eggs as early as 1804, and a Dr. T. Garlick's trout-hatching experiments attracted attention in 1853. U.S. interest received impetus from successes in other countries: By 1850 fish culture was said to be well established in Western Europe, the Balkans, and Scandinavia. Small-scale culture trials took place in Japan as early as 1716, but the Japanese government did not construct field stations for fish propagation until 1876.

Early Public Aquaculture: Extensively Augmenting Wild Stocks

As early as the late 1880s fish stocks in the New England area were showing signs of depletion. Shad, alewife, salmon, and striped bass, which spawned in streams, were among the hardest hit. Reasons for the declines were that fishing was unrestricted and many streams were dammed to satisfy the high demand for hydroelectric power.

In the United States at the turn of the century, nature was criticized as woefully inefficient, and the management of U.S. fisheries was thought to be best done by artificial hatching in protected areas:

> To reap this crop every year, – fishing by effective means must be allowed, and a due proportion must be taken from their spawning grounds so that sufficient ova may be touched by the magic wand of protected propagation to provide for future crops. (Bowers 1898)

The beginning of the U.S. Commission of Fish and Fisheries' movement to reduce depletion of fish stocks was known as the "hatchery solution," and has been credited to several individuals: Seth Green, Dr. Spencer Baird, Dr. Theodatus Garlick, and Livingston Stone, among others. (All have been called either the "father of fish culture" or the "founder of fish culture" in the United States.)

The enthusiasm of fisheries personnel during this period is clear from early commission reports, one of which—*A Manual of Fish-Culture*... (1900)—was very popular. This manual shows that the commission was researching the possible augmentation of wild stocks of a variety of commercial and/or recreational fish species, including alewives, cunner, flounder, golden ide, grayling, haddock, lake herring, mackerel, river herring, pollock, scup, sea bass, sea herring, shad, sheepshead, smelt, Spanish mackerel, tautog, and whitefish. The government eventually cultured seventy-three species of fish, of which forty-seven were freshwater, thirteen marine, and thirteen anadromous.

The myth that marine fish hatcheries could "improve on nature" by protecting early stages of fish and shellfish, thereby increasing commercial and recreational fish catches on an economic basis, seemed very logical at the time. Most biologists and fishery managers were poorly trained, or were political appointees with limited biology backgrounds, few biological data, and little or no experience with marine-fisheries problems. These early fishery workers were amazed at the thousands and even millions of eggs produced by one female fish during a single spawning, let alone annually. However, they were unaware of the extremely high mortality that occurs in nature between the egg stage and adult: For many species, the survival rate is often less than 0.1 percent.

The few examples given in Table 2.1 suffice to illustrate the magnitude of early hatchery production by the U.S. Commission of Fish and Fisheries. The *Fish Hawk,* one of three marine vessels operated by the commission around the turn of the century, was a floating hatchery used to collect eggs, hatch them, and hold the young fish briefly before their release into what appeared to be suitable areas. Most often raised were such species as shad, lobsters, and mackerel. The commission also had five railroad cars especially equipped with tanks and aeration systems for the overland distribution of live fish and eggs.

Whenever fish catches happened to increase after releases, the U.S. hatchery movement got a big boost. Said the Commissioner of Fish and Fisheries in 1887, after the release of 98 million cod fry:

> The unmistakable economic results which have attended these efforts warrant all the time and money devoted to them and justifies the greatest possible expansion of the work.

Hatcheries were held to be the panacea for decreasing wild fish stocks until the 1940s, and the largest proportion of many U.S. fishery budgets went to

Table 2.1. Nineteenth-Century U.S. Government Hatchery Production

Species		Fiscal Year	Number of Fry Planted (millions)
Whitefish		1897–8	88.4
		1898–9	152.7
Grayling		1898–9	4.5
Shad		1897–8	134.5
		1898–9	156.1
Tautog		1896–7	17.6
Flounders		1895–6	8.4
		1896–7	64.1
Cod	"up to and including	1886–97"	449.7

Source: Data from U.S. Commission of Fish and Fisheries, *A Manual of Fish-Culture,* rev. ed., 1900.

maintain them (e. g., for salmon in the west, shad in the east, and whitefish in the Great Lakes). However, stocks declined even in the face of a large hatchery program, and critical examination of hatcheries turned up failure after failure.

More Recent Public Aquaculture

Today there are indeed examples (and in several countries) of biologically and fiscally successful augmentation of natural stocks of commercial and/or recreational fish. These came, however, only after the availability of new information on species biology, the development of productive rearing techniques, the improvement of feeds, and the determination of parameters (e.g., time of year, fish size) for release in the wild. The U.S. National Fish Hatchery System, part of the U.S. Fish and Wildlife Service, presently operates seventy-five facilities, producing hundreds of millions of juvenile fish annually to augment recreational fish stocks. This effort is supported by hatcheries of the National Marine Fisheries Service, others at the state level, and over a thousand in the private sector.

California took a positive step toward determining the value of hatcheries by passing a law in 1983 that provides for the evaluation of artificial propagation, rearing, and stocking of marine fishes. The state hopes that special stamps sold to sportfishermen to fund this program will generate at least $300,000 a year for five or more years. The white sea bass, *Atractoscion nobilis,* and California halibut, *Paralichthys californicus*—popular sport and commercial fish that have been given high priority in the program—have already been reared through all life stages, but economical mass rearing remains to be done. One important aspect of the state program—and one that sets it apart from early hatchery mania

—is that a model is to be constructed, based on knowledge of hatchery economics and species biology, to evaluate and predict probable results of stock enhancements.

Private U.S. Farming

Freshwater Trout

The freshwater, rainbow trout provides a good case study for early private aquacultural trials in the United States. U.S. trout farming started during the 1860s. Aspects of rainbow trout biology were favorable: Females produced large eggs that, together with the sperm from the males, could be stripped from mature fish and mixed together in a bowl. Rates of egg fertilization and fry survival were high, and the young fish even took prepared feeds.

However, despite these favorable biological characteristics, there were deterrents to successful large-scale trout aquaculture. Before freezing was a well-established practice in the fish business, trout could be marketed only seasonally, and marketing strategies for farmed trout were unheard of. Preparation of wet-meal feeds (in the 1940s) was labor intensive, and the supply of raw materials needed for its production was unsteady; moreover, leaching of the feed reduced the water quality in the raceways. Government technical assistance, in such areas as disease control, nutrition, and stock improvement through artificial selection, was almost nonexistent. The market for trout remained weak as late as the early 1960s, not strengthening considerably until later in the decade, when more people became familiar with trout as a result of increased recreational fishing and the improved handling and distribution of frozen fish.

Despite these deterrents, farming attempts increased in number and attracted interest. By 1970 federal research had led to good-quality dry, pelleted feeds containing wheat, fish meal, whey, cottonseed meal, and yeast. Federal studies on artificial selection and disease identification, prevention, and control also greatly aided the industry. State-operated programs for raising trout added further biological knowledge. The spread of such information was encouraged by government agencies and the creation of a trout farmers' association. At last, after a century-long, tedious, and expensive development period, trout farming became successful.

Warm-water Species (Centrarchids, Catfish, and Minnows)

Centrarchids (e.g., sunfish hybrids, crappies, pumpkinseeds, and large- and smallmouth bass) played a major role in initial U.S warm-water pond culture programs. Even early settlers, who constructed ponds for the multiple purposes of power, irrigation, livestock watering, and fire protection, realized that the rearing of freshwater fish would be a valuable additional use. The mid-1930s brought the Dust Bowl, with its severe loss of topsoil across the Great Plains caused by soil depletion and drought. As a result, the development of farm fish ponds as a means of conserving water (and hence topsoil) was encouraged by the U.S. Department of Agriculture's new Soil Conservation Service, which provided technical assistance in their planning and construction.

Much of the early research directed at pond culture of warm-water fishes was conducted by the Illinois Natural History Survey and the Agricultural Experiment Station of the Alabama Polytechnic Institute (now Auburn University). Alabama, with few natural lakes, poor stream fishing, and only a handful of reservoirs, was not well endowed for warm-water sportfishing. This prompted the search for ways to increase warm-water fish production.

The mosquito problem was solved by the discovery that bluegills and small young bass could, under certain conditions (including deepening of the pond edges), be effective in mosquito control. Studies of pond fertility and fertilization methods were required before ponds could become good fish producers. Moreover, the fertility of the ponds often dictated stocking densities.

Stocking schemes caused much disagreement between researchers. Early trials of stocking ponds with a variety of fish species, including large- and smallmouth bass, bluegill, crappies, and catfishes, did not yield clear-cut data.

Good results were obtained by stocking a combination of bluegill and largemouth bass. The concept of a balance in numbers between the two species was put forth by researchers H. S. Swingle and E. V. Smith (Auburn University). When in the proper ratio, both species were able to reproduce, grow well, and yield good harvestable crops; but an overabundance of one species or the other did not permit good harvests.

Stocking ratios, however, could not be used as hard and fast rules: Their effectiveness, it was found, could be determined by local pond conditions. In certain ponds, once a suitable balance was obtained, it tended to remain for a considerable time—sometimes for as long as several decades.

The number of fish ponds in the United States increased from about twenty thousand in 1936 to over two million in 1965. Federal and state hatcheries provided fingerlings to stock these ponds. Later, when it was learned that channel catfish could be induced to spawn in standing water, interest developed in catfish culture. The commercial production of cultured channel catfish has been one of the best examples of successful U.S. aquaculture ventures: In 1961, about fifty thousand pounds of catfish were marketed; by 1991 over 178,000 metric tons (mt) were being produced (live weight).

As a result of increases in sportfishing, sparked by the public having more leisure time and more ponds and reservoirs open to them, bait minnows became very much in demand and difficult to find. When it seemed that trapping and seining in natural waters could not satisfy the demand for live bait, commercial production methods were described and information published by the U.S. Fish and Wildlife Service. Many of the biological aspects of the minnow species were not well known, and ignorance of the role of parasites and diseases was a stumbling block in many early trials. Methods of effective pond culture that had been trade secrets were made known, and with newly developed feeds and fertilizers, production increased dramatically.

Future of Aquaculture

In sharp contrast to the many dismal failures it experienced over the past three decades, U.S. aquaculture has recently made considerable strides in production (see Table 2.2); moreover, continued increases in production are predicted. By taking a more deliberate, measured approach to aquacultural operations, and incorporating ongoing scientific and technological inputs, U.S. aquaculture has approached the dawn of a new era.

Serious barriers to the aquacultural desirability of some species of finfish have been removed through genetic engineering. For instance, crossing striped bass (*Morone saxatilis*) with white bass (*M. chrysops*) has yielded improved growth, greater disease resistance, and better survival. Hybridization of sunfish (bluegills) in the genus *Lepomis* have successfully produced larger fish. Also, grass carp (or white amur, *Ctenopharyngodon idella*), notorious for competing with local species for food and shelter, have become widely used for aquatic weed control after having been sterilized by a technique known as *polyploidy*.

Transgenic engineering ("gene splicing") involves the transfer into one fish of genes cloned from another genus or species. The cloned material is injected into an egg of the receptor species; if the transfer is successful, the gene remains in the fish's lineage. Depending on the genes transferred, this technique can improve growth, resistance to cold, immunity to disease, and so on. However, moving transgenic engineering out of the laboratory and applying it to commercial aquaculture is problematic and a hotly contested issue. Outspoken ecologists foresee "superfish" escaping from labs and fish farms, becoming established in the wild, and upsetting if not destroying the local ecology. Nevertheless, some scientists feel that gene splicing will be a routine part of commercial aquaculture by the year 2000.

Technological advances notwithstanding, there are still important constraints on the successful culture of many species—legalities and permits, environmental degradation, disease control, human health, and competition for resources. Federal support and cooperation have declined, with the burden passed to state agencies and the private sector. Aquacultural production has been boosted, however, by the provision of technical assistance in the thirty-two states that have aquaculture coordinators (see Appendix E).

U. S. Aquacultural Production by Major Species Groups

Group	Production (mt)[a]
Finfish	262,149
Mollusks	186,664
Crustaceans	80,863
Seaweed	210
Total	529,886[b]

[a]1986 FAO data (metric tons); FAO lists Pacific salmon, catfish, and mollusks as key (*Aquaculture Minutes*, no. 8, August 6, 1990).
[b]Comprises 4% of total 1986 world aquacultural production of 12,103,418 mt.

Finfishes

The channel catfish (*Ictalurus punctatus*), whose sales were once restricted to the southern states, has substantially broadened its base to emerge as one of the major U.S. freshwater aquacultural species. Roughly 2,500 mt were produced in 1970; but with the 1985 introduction of catfish in fast-food restaurants, output skyrocketed, with over 163,000 mt produced in 1991. These increases have begun to taper off due to falling farm gate prices, and product identification and advertising are on the rise: It seems that industry volume is stabilizing and may have to contend with price adjustments. Other factors influencing the future of the catfish industry include improved quality control (federal inspection), product diversification (from whole fish to fillets), and the consolidation of the processing sector (four companies now process over 60% of all U.S. catfish).

Of the salmonid fishes, trout remains a good market species, with generally stable pricing and moderate industrial growth. Because the number of desirable sites are limited by the requirement for abundant, high-quality water, the improvement of production is being sought by technological means rather than expansion. Results to date in the U.S. net-pen salmon industry, on the other hand, do not suggest that the United States can become a serious player in the world salmon market against the well-established (and/or subsidized) net-pen industries of Norway, Iceland, Scotland, Canada, and Chile. Recent scientific inquiries into salmon ranching and the condition of wild salmon question the advisability of continuing these practices on a large scale. (See also the section "Salmonids" in Chapter 3.)

Tilapia production, limited geographically by the requirement for warm water, has continued to increase at about 5 percent per year, coming almost exclusively from Florida and California. Thanks to hybridization, nearly all tilapia produced are the faster-growing males. Consumer acceptance is increasing with improvements in the fish's color and its recent appearance in supermarkets.

Farmed striped bass hybrids enjoyed a favorable market when wild stocks were low; the number of farms jumped and prices dropped. Considerable fluctuations in price are expected as hybridizers continue to compete with fisheries.

Crustaceans

Although South Carolina (summer crop only) and Texas did produce two million pounds (heads on) of so-called warm-water shrimp in 1990, culture of this genus (*Penaeus*) does not hold much hope for U.S. investors. (See "Shrimps and Prawns" in Chapter 3.) However, the fast-growing giant river prawn (*Macrobrachium rosenbergii*), of minor commercial value to fisheries, has farming potential where waters are sufficiently warm. Hawaii has been successful with this species, selling it on the U.S. mainland. As an exotic, it does have some marketing constraints, and strong competition from well-known marine shrimp is a big hurdle.

U.S. crayfish production has reached the big time: In 1990 seventy-one million pounds were produced, mainly in Louisiana. Prices are higher for the soft-shelled (newly molted) variety. Crayfish can be made available throughout the year by freezing.

Mollusks

There have been few changes in oyster culture techniques. Most of the output comes from public grounds, where some quasi-aquacultural methods are practiced, such as spreading oyster shells (*cultch*) to collect wild spat. Private oyster farming on leased sea bottoms with protection from predators by off-the-bottom culture is expected to continue because of the higher returns. Oyster culture in the Pacific Northwest involves intensive techniques, including the use of hatcheries. "Remote setting" is now being used on the Pacific Coast of North America: Hatchery larvae ready for final settlement are shipped to selected locations, where they will metamorphose and set on cultch in tanks; the cultch can then be placed in oyster beds. This solves many of the difficulties of obtaining a good setting of wild spat.

The recent and increased occurrences of human disease resulting from dinoflagellate toxins in mussels (paralytic shellfish poisoning; see "Mussels" in Chapter 3), added to the high cost of farming, does not suggest a major increase in U.S. aquaculture of this relatively low-demand shellfish. Likewise, clam culture, although increasing in popularity, suffers from many constraints —environmental restrictions, financial limitations, and needed research—and does not appear to have a bright future in the United States.

REFERENCES

Bowen, J. T. 1970. A history of fish culture as related to the development of fishery programs. In *A century of fisheries in North America,* ed. N. G. Benson, pp. 71–93. American Fisheries Society special publ. no. 7. Washington, DC: American Fisheries Society.

Bowers, G. M. 1898. *Report of the Commissioner for the year ending June 30, 1897.* Washington, DC: U.S. Commission of Fish and Fisheries, Bureau of Fisheries.

Swingle, H. S. 1970. History of warmwater pond culture in the United States. In *A century of fisheries in North America,* ed. N. G. Benson, pp. 95–105. American Fisheries Society special publ. no. 7. Washington, DC: American Fisheries Society.

Tiddens, A. 1990. Aquaculture in America: The role of science, government and the entrepreneur. Boulder, CO: Westview Press.

Titcomb, J. W. 1908. Fish-cultural practices in the United States Bureau of Fisheries. *Bulletin of the Bureau of Fisheries* 28(Pt. 2):697–757.

U.S. Commission of Fish and Fisheries. 1900. *A manual of fish-culture based on the methods of the United States Commission of Fish and Fisheries, with chapters on the cultivation of oysters and frogs.* Rev. ed. Washington, DC: U.S. Government Printing Office.

U.S. Commission of Fish and Fisheries—Bureau of Fisheries. 1873–1949. 24 vols. *Report of the Commissioner for 1871/72–1882, 1884, 1887–1888, 1892, 1897, 1899, 1911, 1923–1924, 1928, 1931–1932, 1937–1939.* Washington, DC: U.S. Government Printing Office.

Wood, E. M. 1953. A century of American fish culture, 1853–1953. *Progressive Fish-Culturist* 15(4):147–62.

CANADIAN AQUACULTURE

Aquacultural History

The origin and development of aquaculture in Canada parallel those of the United States. It began during the 1850s, when brook trout and Atlantic salmon were artificially spawned. This resulted in the construction and operation of federal hatcheries essentially from coast to coast, with salmon hatcheries on the coasts and trout inland. The first hatchery for raising Pacific salmon was constructed on the Frazer River (western Canada) in 1883. By 1950 these federal hatcheries were producing in excess of three-quarters of a billion freshwater and anadromous fish annually. Private aquaculture started with salmon in the late nineteenth century, during which time several commercial hatcheries were constructed.

"Jar culture" started in 1868, when semibuoyant freshwater whitefish eggs were incubated. A commercial hatchery using this technique was set up on the Detroit River in 1875. A series of such hatcheries were built on the Great Lakes and later in western Canada, producing and distributing to Canadian waters millions of eggs and fry of whitefish, pickerel (walleye), lake herring, and perch. As in the United States, hatchery programs were fueled by those wanting improved recreational fishing and enhancement of commercial fish stocks;

and, as in the United States, the futility of trying to increase the abundance of wild stocks finally became apparent.

Warm-water fish culture efforts began in 1872 and included the freshwater black bass (*Micropterus* spp.). Carp invaded Canadian fresh waters and thrived despite the cold water temperatures.

In more than a hundred years of Canadian aquaculture, government hatcheries have raised sport fish for stocking public waters (sometimes called "public aquaculture"), as well as private. At least twenty-eight native species of finfish and seven exotic species have been cultured in Canada during this time. The rearing of fish for human consumption in Canada is rather new, however. Recent endeavors include oyster, mussel, and finfish farming.

Little attention has been directed to commercial aquaculture of the American oyster, *Crassostrea virginica,* on Canada's Pacific Coast, where the waters are generally too cold for the species to spawn and population levels are consequently low. On Canada's eastern coast, however, this oyster was considered for aquaculture as early as 1865, and over the years American-oyster culture leases were available in Nova Scotia and Prince Edward Island. Production rose until about 1940 and then declined.

Because the oyster native to the Canadian Pacific Coast, the Olympia oyster (*Ostrea lurida*), is by reason of its small size and slow growth not suitable for aquaculture, the Canadians, like the Americans, imported the large Pacific oyster (also called the Japanese oyster), *Crassostrea gigas*. Although the Pacific oyster still does not spawn successfully to any degree in Canadian waters, it does grow and survive well. Conditions suitable for its reproduction do exist in one western coastal area, Pendrell Sound.

The first Atlantic salmon farm in Nova Scotia failed in 1972, about three years after it was started; however, an experimental salmon farm was started on the western coast of Canada in 1974 at the Pacific Biological Station, Nanaimo, British Columbia. There, several species of Pacific salmon and steelhead trout were tested for their suitability for commercial culture. This demonstration unit hatched and reared fish in fresh water that was heated to accelerate growth of the young. They were then placed in salt water in net pens to be raised to commercial sizes. A 1989 estimate shows that there are about 45 salmon farms on the Atlantic Coast (primarily in the Bay of Fundy) and about 150 on the Pacific. The majority of salmon farmers rely on government or private salmon hatcheries to supply them with smolts to grow to market size.

Trout farming was purported in 1970 as having the potential of a multi-million-dollar industry in the prairie land of Manitoba, Alberta, and Saskatchewan, where there are many potholes (actually, small lakes and sloughs). Farmers had tried unsuccessfully to drain these potholes so that the land could be used for agricultural crops. With a program initiated by the Freshwater Institute at Winnipeg, rainbow trout (*Oncorhynchus mykiss* [formerly *Salmo gairdneri*])

were stocked as fry in the spring and could be harvested in the au
they reached an average size of about eleven inches.

Though some trout farmers provide supplemental feed, natural f
is abundant in these water bodies, mainly in the form of a small shrimplike freshwater crustacean, *Gammarus lacustris*. Other advantages of this scheme are that the trout grow and survive well during the warm summers, whereas the number of predators on trout in the pothole lakes is controlled by winter freezes. The estimated value for trout produced was about $75 per acre, and the net return to the farmer is estimated to average well above the $25 per acre obtained from a normal harvest of wheat. In recent years, "borrow pits" (from which soil is removed for filling low areas) have been converted to fish ponds.

There are some disadvantages to pothole-trout farming, however. Oxygen levels may be reduced to dangerously low levels by a heavy algal bloom. Algae may also impart a muddy or off-flavor to the flesh, the correction of which may require holding the fish in algae-free water prior to marketing. (This same off-flavor occasionally occurs in southern U.S. freshwater catfish and in South American warm-water shrimp.) Also, since many of these potholes are rather shallow, predatory birds can reduce trout production.

Credit is given to Canada for starting the first eel culture station in North America, located in New Brunswick. The station, which began operation in the mid-1970s, boasted a heated spawning pond and indoor rearing ponds. The eels raised in this operation were judged to be of good quality for eating, with soft skin and tender flesh. Fishermen in the New Brunswick area were pleased with the project because it tended to control the eel population in the rivers, thereby reducing eel predation on salmon and trout. However, trials suggested that the station was not economical; in fact, a recent assessment of the potential for eel aquaculture rates this species as very low.

Current Status of Aquaculture

Species whose culture can now be pursued vigorously on a commercial scale include chinook, coho, and sockeye salmon, rainbow and brook trout, and Atlantic and Pacific oysters.

Salmon

A relatively new program for salmon focuses not on the spawning, hatching, and rearing of young in indoor artificial habitats, but rather on the improvement of natural spawning habitats, in part by the use of spawning channels and stream incubation boxes. *Spawning channels* increase production by providing spawning gravel of the proper size, suitable current velocity, and an optimum number of spawners per unit area, as well as by excluding predators. They have been immensely successful in the Frazer River on Canada's Pacific Coast, where five channels, built during 1969–74 at a cost $1.7 million, have yielded sockeye and pink salmon with a landed value of over $10 million. *Stream incubation boxes* are simply open-top mesh boxes in which fertilized eggs are placed with gravel of a size that permits adequate water flow; the boxes are then buried in areas with substrate current velocities that will ensure high egg and fry survival.

Aerial spraying of aqueous solutions of inorganic nitrogen and phosphorus can increase production of young sockeye salmon in nutrient-deficient lakes. Nutrient enrichment increases phyto- and zooplankton production, thus providing more food during the lake-dwelling period of the sockeye. In British Columbia, smolts leaving most fertilized lakes were larger than in pretreatment condition, and observed marine survival of returning adults surpasses anticipated levels.

Salmon farming in British Columbia took a giant leap forward during 1985–9, going from approximately 107 mt to about 13,600 mt. Unfortunately, a 1989 salmon price collapse greatly reduced production, and belt-tightening and more competitive and efficient business practices were begun in the estimated two-thirds of the companies still in business. The farmed salmon industry on the Atlantic Coast weathered the price collapse: Production increased and was estimated to be about 17,000 mt for 1990.

Freshwater Fish

In addition to the salmon released in commercial and recreational fisheries, millions of freshwater fish—cisco, whitefish, pickerel, pike, perch, black bass, and maskinonge—are raised in federal, provincial, and private hatcheries for live release. This artificial propagation continues despite the low survival of many of the released fish.

Private freshwater aquaculture in Canada is essentially the culture of rainbow trout. Canadian farms have recently produced over 10 million pounds of freshwater fish annually. Of this, about 72 percent was sold for human consumption. Trout were also sold as fry, fingerlings, and yearlings to stock private waters.

Invertebrates

Oysters are the only invertebrate marine species contributing to aquaculture production on both coasts. Governmental assistance for oyster farmers has been in the form of federally funded laboratories that studied the life history and rearing techniques of this mollusk on both coasts. Their results are well described in excellent government publications available from Information Canada.

Canadian blue mussel farming has had its share of problems. As if cold weather and harvest-hindering ice were not enough of a headache, there have been social and economic difficulties as well. Mussel longlines are considered offensive to some water users, especially traditional capture fishermen, who make their complaints known. Another more serious and unforeseen problem arose on Prince Edward Island with the bloom of a toxic diatom, causing the accumulation of domoic acid in mussels in several farms. The poisoning of 129 persons and death of two were a serious setback to the industry, leading to loss of consumer confidence. Production monitoring and a public relations campaign reestablished the market by 1989, with Canadian blue mussel farmers on the Atlantic Coast turning in production figures of more than 2,720 mt, worth over (U.S.)$5.5 million. About 78 percent of this tonnage came from Prince Edward Island; the remainder came from Nova Scotia, Quebec, Newfoundland, and New Brunswick, in descending order. Blue mussel trials on the coast of British Columbia have been beset with such difficulties as the competition for food and areas of attachment by fouling organisms and the natural mortality caused by sea birds.

After about 20 years of aquacultural research and development trials on the eastern coast of Canada, the commercial aspects of scallop farming, using the bay scallop and three other species, are not encouraging. The nonnative bay scallop, *Argopecten irradians,* seems to be doing well in Nova Scotia, where it has undergone trials for aquacultural potential. On the western coast, the rock (or purple-hinge) scallop, *Crassadoma gigantea* (= *Hinnites multirugosus*), which attaches itself to a clean, firm surface, has been suggested as a viable candidate.

Although most of the early biological constraints to scallop farming have been overcome, obvious economic hurdles must be cleared before scallop farming can be successful in Canada. Production costs are high, with bottom culture apparently the only economical method, and there is stiff competition from the capture fisheries, as well as from foreign imports. The Japanese scallop (*Patinopecten yessoensis*) is being researched on Canada's Pacific Coast; other species (the pink scallops and weathervane scallops) have been undergoing trials involving lantern nets in the Strait of Georgia.

Outlook for the Future

Many of the requirements for successful aquacultural ventures already exist in Canada. Cool-water salmonids are naturally abundant, and there is potential for other marine species. The protected bays, inlets, and straits of the Pacific Coast offer a most suitable environment for both net-pen rearing of salmon and oyster production. Moreover, freshwater aquaculture products do not require market promotion because traditional Canadian markets are in place. However, more technical information is still needed to minimize risk in the form of biological economics. One major limitation to the development of successful aquaculture would seem to be underfinancing: Cash resources are frequently used up before profitable production begins. For many years Canadian aquaculturists have received no appreciable governmental assistance for either financing, advertising, or marketing. In addition, many legal aspects restrict the purchase, transport, and sale of privately raised fish and shellfish.

Canadian oyster culture appears to be troubled by a lack of solid business management and by not employing the most recent technology. It has been said that if the results of government education programs were used wisely, oyster production could be greatly increased. The domestic market for oyster consumers is approximately four times domestic production; presently, imports from the United States and Japan make up the difference.

Species other than those presently farmed may merit large-scale production after determination of market demand and product quality. These include pink, chum, and Atlantic salmon, sablefish, eels, and freshwater crayfish. Species recommended for production on a small scale (at least until more information is available) are flatfish, walleye, whitefish, clams, quahogs, mussels, scallops, lobsters, and seaweeds.

The "stars" of Canadian aquaculture are species farmed on the Atlantic Coast: the blue mussels and the Bay of Fundy salmon. A 1989 report prepared for Canada's Minister of Fisheries and Oceans forecasts farmed salmon production of at least 31,000 tons by the year 2000, with a best-case prediction of 66,000 tons. There may also in future be considerable improvement in the production of pothole trout, a rising star.

REFERENCES

Boghen, A. D., ed. 1989. *Cold-water aquaculture in Atlantic Canada.* Moncton, NB: Canadian Institute for Research on Regional Development.

"Focus on Canada." 1990. *World Aquaculture* 21(2):1–100.

MacCrimmon, H. R. 1983. Canada. In *World fish farming: Cultivation and economics,* ed. E. E. Brown, pp. 67–81. 2nd ed. Westport, CT: AVI Publishing Co.

MacCrimmon, H. R., J. E. Stewart, and J. R. Brett. 1974. *Aquaculture in Canada: The practice and the promise.* Bulletin of the Fisheries Research Board of Canada 188. Ottawa: Department of the Environment, Fisheries and Marine Service.

Manci, W. E. 1990. Canada: Ready for the '90s. *Aquaculture Magazine* 16(6):44–9.

MEXICAN AQUACULTURE

Geography and Habitats

Mexico has only one large river, the Rio Grande, and two great mountain chains, one to the east and the other to the west. They meet to the south in a mountainous region that is the beginning of the plateau of Central America. These mountains and mountain ranges, most of which are of volcanic origin, provide more rainfall and a cooler temperature range than is found in the lowlands, thereby improving the potential for aquaculture. These upland cold areas have average temperatures of 59–63 °F, and trout (*Oncorhynchus mykiss*) do well in these conditions; however, there are no large parcels of land suited for extensive aquaculture in these areas. Production levels probably do not exceed those required for local consumption.

The total area of Mexican lowlands with fresh water is large (estimated at over 1.2 million acres), with many rivers and streams that have aquacultural potential. Species suitable for lowland freshwater areas are tilapia, carp, catfish, mullet, river shrimp, and ornamental fish. Seed from fish hatcheries has been used to restock large water bodies.

The greatest potential seems to lie in the coastal lagoons and estuaries, where salinities range from brackish water to ocean seawater. The total area of these lagoons is estimated at 2.3 million acres. Species suitable for rearing in such lagoons include mollusks (oysters, clams, scallops, mussels, pen shells, abalone) and marine shrimp; but despite the large number of molluscan species, only oyster culture has been actively pursued. These coastal areas seem to have all the requirements for increased aquacultural production.

Improvement of the water quality in the costal lagoons came under consideration with the formation in 1970 of a governmental agency, the Direccíon General de Acuacultura. Targeted were those areas whose water quality had been lowered either by natural causes or by human impact on the environment. This organization was followed in 1976 by a fisheries department, whose policy was to research species for introduction into aquaculture.

Status of Aquaculture

The diversity of habitats has yielded a diversity of species on the northwest coast, where five species of abalone and one bivalve are all fished intensively, with another six species of gastropod or bivalve fished occasionally or locally. Of the species present, only three have been used for aquaculture, all of them oysters: the native oyster, *Crassostrea corteziensis;* the rock oyster, *C. iridescens;* and mother-of-pearl, *Pinctada mazatlanica.* As in Canada and the United States, Mexico also introduced the large Japanese oyster, *C. gigas.* The native oyster, traditionally collected for local use, gradually entered commercial markets. Culture techniques were under study for mother-of-pearl oysters as early as 1910, but this trial ended in failure due to mass mortality of the oysters. As fishing intensity increased (and therefore landings decreased), further attempts were made to culture mollusks. Abalone were cultured for private use and to augment wild stocks. In addition the pen shell (*Pinna rugosa*), mother-of-pearl (*Pinctada mazatlanica*), and bay scallop (*Argopecten circularis*) were cultured on the western coast.

Despite Mexico's numerous molluscan species—the Pacific Coast has at least sixty-seven species—aquacultural research has been directed at fewer than ten of these. The following are believed to have been important deterrents to the pursuit of molluscan aquaculture: lack of suitable scientific data and aquacultural trials for some species, insufficient control of water quality, and the idea of paying for what has always been a common property resource that could be taken for free.

Others blame the low level of aquacultural development on Mexican fisheries legislation. The following is from an Inter-American Development Bank (1977) report entitled *Aquaculture in Latin America:*

> The Federal Law on the Development of Fishing (May 10, 1972) has no major provisions regarding aquaculture. Its main contribution in this regard was to establish the *ejido* [public or common land] fishing cooperatives, permitting the development of aquaculture in rural communities. In accordance with the Law on Fisheries, the principal native species suitable for culture and of greatest economic value—such as shrimp, oysters, lobster, sea turtles and clams—may be exploited or cultivated only by cooperatives and their production must be sold to State companies. This limits the action to private individuals in brackish water aquaculture. Under the Constitution (Article 27), foreign individuals or companies may not own or obtain concessions to land or waters, which precludes the participation of non-Mexican companies.

Private businesses had little interest in investing in brackish-water species with poor market potential and few if any aquacultural trials. The purpose behind excluding businessmen from the "prime" species—shrimp, oysters, and so on—had been to allow poor families to make a living farming them (and other species) through the formation of government cooperatives. Rules for establishing and running such cooperatives have been formalized. Once a proposed cooperative receives government approval, it can apply for federal assistance in the form of low-interest-bearing loans from a specialized bank (BANPESCA). Assistance with construction of ponds, some types of equipment, and technical information can also come from the government. When sales begin, loan interest and other costs are paid off the top; a portion of the remainder of the income is then divided among the members of the cooperative.

Naturally, "work now for possible pay later" is hard for poor families to contend with when they are living hand-to-mouth and there is no food on the table. Also, although the concept behind the cooperatives is truly noble, the poor families that are supposed to be helped by the system complain that little of the income filters down to them. This may be partly because cooperatives have too many members to provide a realistic return on effort expended by the members. Proponents of cooperatives point out that, in some types of aquacultural cooperative, operations require only minimal labor input of its members; for instance, oyster farming usually has only two labor-intensive periods during the year. However, as with any aquaculture venture, there can be emergencies that require maximum effort from all available labor.

Clams

Few of the many species of Mexican mollusks have been considered for culture generally because of a lack of biological information. The pen shell (*Pinna rugosa*) and some species of scallops and clams are candidates with potential for commercial culture. Like the "cherry stone" clams sold in the United States, these Mexican clam species can be marketed at small size for a price possibly higher than could be realized were they raised to full-sized adults.

Oysters

Polluted water and oily taste of the product has doomed oyster aquaculture in some areas of the Gulf Coast, where the native oyster (*Crassostrea corteziensis*) has suffered at the hands of the oil industry. Today, this oyster is farmed using Japanese raft culture on the western coast.

Freshwater Prawns

In 1971 a small group of private developers made preliminary rearing and culture trials with the freshwater prawn *Macrobrachium americanum,* a native of the Pacific Coast of Mexico. It was given serious consideration for commercial aquaculture because of the large size it reached in nature (about 1 lb for males and 8 oz for females), its large, robust and edible claws, and a favorable tail–head ratio. Large gravid females were easily obtained in nearby rivers and streams for hatching and rearing. Since no prepared food was available for this species, they were fed scraps from the tables of local people.

The freshwater prawn *M. rosenbergii* was transplanted to Mexico from California (in 1973) and Hawaii (in 1974). In 1975 over 14,000 tons of juvenile freshwater prawns were produced in Mexico.

Marine Shrimp

Several species of marine shrimp are being raised: the Pacific white shrimp (*Penaeus vannamei*), the blue shrimp (*P. stylirostris*), and the yellowleg shrimp (*P. californiensis*). Without the use of hatcheries, it is necessary either to collect postlarval shrimp in the wild or purchase them from artisanal fishermen in order to seed grow-out ponds to produce market-sized shrimp.

Using wild-caught seed (postlarvae) for shrimp farms is commonly practiced in Mexico because it requires less skill and expense than the mating of shrimp and raising of early stages in captivity. Since the abundance of wild seed relies on the vicissitudes of nature, its availability is highly variable from year to year.

Of course, a much greater production per acre can be realized given maturation facilities capable of maturing, mating, and spawning adult shrimp, and a hatchery to mass-rear pure cultures of young shrimp; but shrimp maturation and hatchery operations are very expensive to construct and maintain. Mexican cooperatives normally have neither the funds to include these technologies in their shrimp farms nor the skilled workers to run such facilities. A production hatchery must have rather a large and dedicated work force—at times, seven days a week, around the clock.

Mexico is faced with the same problems as other Central and South American shrimp farmers. Shrimp make a good meal for many predators, including birds, fishes, and crabs. Fish and crab predators are especially bad when the seed shrimp for pond grow-out is obtained from the wild: Numerous larval-stage predatory fishes can accidentally be introduced into the ponds and, when large enough, will feed heavily on the shrimp.

Several strains of brine shrimp are also known from southeastern Mexico. The biology of these strains is being studied because they are important as food not only in the cultivation of early stages of penaeid shrimp larvae, but also for flamingos in government-operated refuges.

Finfish

Biological studies of the sea catfish, *Galeichthys caerulescens,* suggest that it may be a good candidate for culture in the numerous Mexican lagoons. This fish can tolerate temperatures of 66–95 °F and salinities of 0–45 ppt (parts per thousand). The lack of predators on the sea catfish is a plus for its culture.

Wild carp, which may reach a length of up to three feet and weigh thirty pounds, are good species for aquaculture, especially in warm-water habitats. They eat natural foods and tolerate high density in ponds; they spawn readily in captivity, and can be induced to spawn more frequently by injecting an extract of carp pituitary gland into mature fish. The ease of carp culture, which involves little financial investment and low-level technology, makes this species desirable for many rural Mexican areas. Approximately 6,000 mt of carp were produced in Mexico in 1975.

Both rainbow trout (*Oncorhynchus mykiss*) and brook trout (*Savelinus fontinalis*) are raised in Mexican government hatcheries located in the mountains—some about 9,000 feet above sea level—to obtain conditions suitable for rearing. Springs provide water of excellent quality, which is gravity-fed to the trout hatcheries and farms. Male trout reach maturity at age 2–6 and females at age 3–6. Commercial farms selling trout for human food average about 25 mt/acre. The juvenile trout produced are shipped to various locations in Mexico, where they are used for stocking rivers, lakes, and reservoirs for sportfishing.

Future of Aquaculture

As mentioned above, for many years aquaculture investment in Mexico for the "prime" aquaculture species was restricted to cooperatives and *ejidos;* however, plans were made early in 1989 to step up promotion of private investment in Mexican aquaculture. In December of that year, Mexico passed legislation permitting private ventures in shrimp farming and allowing foreign investment with a Mexican partner. This law, which became effective April 1, 1990, is expected to increase substantially the flow of investment funds badly needed for shrimp aquaculture. For example, in the state of Sinaloa, midway down the Pacific Coast of Mexico, there are now some ninety-six shrimp farms; at least eighty of these are now joint ventures with private investment. This recent economic restructuring has changed the operation of these farms and is expected to increase their production. With a new, large, and well-funded venture moving into the area (startup date 1992), this hub of Mexican shrimp farming industry may reach 30,000 mt available for harvest during 1994. Mexican shrimp farm production overall is expected to increase significantly, and perhaps even outstrip that of Ecuador. This forecast is based both on the enormous amount of Mexican land available that is suitable for shrimp production, and on a plan to construct hatcheries to produce young shrimp where farmers do not have access to natural waters containing larval shrimp.

Mexico has come to realize that the general lack of information on the biology, life histories, and aquacultural suitability of many indigenous species has limited production. Steps have been taken to correct this situation, and should result in greater production. For instance, in 1989 a national committee was established to coordinate and monitor the country's by then twenty-five or so institutions devoted to basic and applied research bearing on aquaculture (Noriega-Curtis and Rivas, 1989). Also, to assist farmers interested in producing fish for private use or local sales, more juvenile fish are being raised in government hatcheries for distribution and for stocking farm ponds.

High inflation rates in Mexico have harmed many businesses, and aquaculture is no exception; but inflation dropped from 57 percent in 1986 to 20 percent in 1990. Still, even in 1986 Mexican aquaculture produced 43,210 mt of aquatic species, equal to 43 percent of Latin American production by weight in that year. Although reliable current data on total Mexican aquacultural output are unavailable, the industry does appear to be making headway, and its success has encouraged more aquacultural ventures. Thus, though many labor-related problems and certain restrictions on private investment remain, further reductions in the inflation rate and a more stable economy would undoubtedly stimulate aquacultural production.

REFERENCES

Conrad, J. 1985. Mexico's cooperative oyster and shrimp farms. *Aquaculture Magazine* 11(5): 46–9.

Inter-American Development Bank. 1977. *Aquaculture in Latin America.* Washington, DC: Inter-American Development Bank.

Noriega-Curtis, P., and J. V. Rivas. 1989. *A regional survey of the aquaculture sector in Latin America.* Aquaculture Development and Coordination Programme ADCP/REP/89/93. Rome: United Nations Development Programme, FAO.

Pedini Fernando-Criado, M., ed. 1984. *Informes nacionales sobre el desarrollo de la acuicultura en America Latina.* FAO informe de pesca no. 294, suplemento. Rome: FAO.

3

Species for Human Food

PLANTS

Vegetable Hydroponics

Hydroponics is the growing of plants in nutrient solutions, with or without an inert medium to provide mechanical support. Vegetables normally raised in soil can be grown in a solution or moist inert material that contains the necessary nutrients. The hydroponic techniques used to raise such vegetables as cabbage, lettuce, and tomatoes can be integrated with fish farming: The plants take needed nutrients from wastewater from the fish-rearing facility, thereby often improving water quality. Tilapia have shown great improvements in feed conversion, and hence greater production of marketable fish, when their farming was integrated with hydroponics, due to the improved water quality in the integrated system.

There are some drawbacks to such integration: Special tanks are required; sludge and toxic salt levels must be controlled; fertilizers must still be added to ensure good nutrient balance and plant production; and pesticides used on the plants must be monitored for toxicity to the fish and accumulation in their flesh. Nevertheless, the vegetable hydroponics–fish farming system is recommended for areas where *meager supplies of fresh water* limit both aquacultural and agricultural food production. The success of trial studies on St. Croix suggests that large-scale commercial production of both fish and vegetables can be achieved.

REFERENCE

Rakocy, J. E. 1989. Vegetable hydroponics and fish culture: A productive interface. *World Aquaculture* 20(3):42–7.

Algae

This section gives examples only of algae used directly as human food; non-food macroalgae cultured for chemical extraction are discussed in Chapter 4.

Microalgae

Microalgae (especially *Spirulina*) have been farmed for human food, usually as a protein or nutrition supplement; but Americans tend not to like the taste of these minute plants.

Macroalgae

Many species of macroalgae occur along U.S. seacoasts, mostly in intertidal zones. In Canada, about 300 species grow along the Atlantic Coast. Of the numerous species found in Hawaiian waters, over 70 are edible. At least seven species have been reared in experimental or pilot-scale operations, with cultivation for human food most successful in Asia. Most macroalgae contain minerals and vitamins of value to the human body: Dulse, a species of coarse red seaweed, contains protein (25%), carbohydrates (44%), and mineral salts (27%).

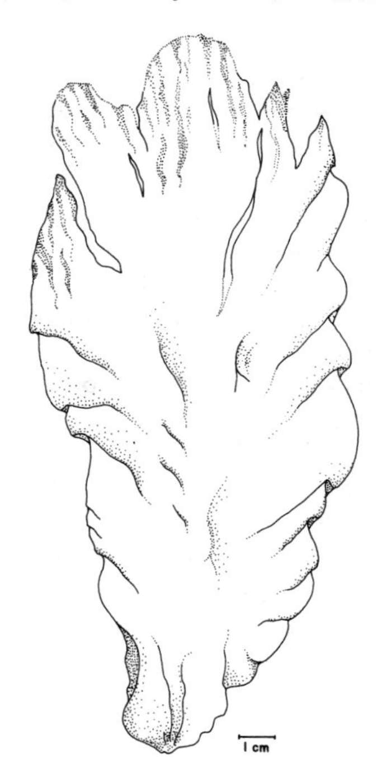

SCIENTIFIC AND COMMON NAMES: ***Porphyra spp.*, Nori**
Several of the numerous species known from this genus are cultivated.

DESCRIPTION AND DISTINCTIVE CHARACTERISTICS: Fronds are either long, narrow, and knifelike, or broad and fan-shaped with ruffled edges. Color varies from purple to deep rose to black.

HABITAT RANGE: Intertidal zone, in temperate waters of the Northern Hemisphere.

REPRODUCTION AND DEVELOPMENT: Reproduction is sexual: When water temperature reaches about 72 °F, plants release spores (called *conchocelis*) that bore into oyster shells. Upon leaving the oyster shell, conchocelis find attachment and grow to large plants.

AGE AND GROWTH: Blades may reach 3 feet in length and 1 foot wide in about 7–8 months.

FOOD AND FEEDING: Photosynthesis; extracts nutrients from shallow water.

PARASITES AND DISEASE: Red rot, a fungal disease.

PREDATORS AND COMPETITORS: Predators include sea urchins and, in culture, snails, isopods, and amphipods.

AQUACULTURAL POTENTIAL: A traditional food in Asia, where it is extensively cultured, especially in Japan. Some is eaten fresh, but most is sun-dried before sale. The market in North America is weak. Due to high levels of pollution in Japan's coastal waters and increased demand for nori, foreign sources are being sought.

REGIONS WHERE FARMED AND/OR RESEARCHED: Farmed in Tokyo Bay, Japan, since the late 17th century; Hawaii (Kona), Washington, Alaska, North Carolina, Oregon; British Columbia and Prince Edward Island, Canada; China, Korea, Philippines, Norway, England, and Chile.

REFERENCES

Bird, K. T., and P. H. Benson, eds. 1987. *Seaweed cultivation for renewable resources.* Developments in Aquaculture and Fisheries Science, vol. 16. Amsterdam: Elsevier.

Hansen, J. E., J. E. Packard, and W. T. Doyle. 1981. *Mariculture of red seaweeds.* California Sea Grant report T-CSGCP-002. La Jolla: California Sea Grant.

Mathieson, A. C. 1982. Seaweed cultivation: A review. In *Proceedings of the Sixth U.S.–Japan Meeting on Aquaculture, Santa Barbara, California, August 27–28, 1977,* ed. C. J. Sindermann, pp. 25–66. NOAA technical report NMFS-CIRC 442. Seattle: National Marine Fisheries Service.

INVERTEBRATES

These animals, low on the evolutionary scale, are simple in structure and range in size from microscopic (protozoans) to 10 ft across at the legs (king crab), to about 50 ft in total length (giant squid). Many are asymmetrical; others are spherically, radially, or bilaterally symmetrical. More advanced forms of invertebrate have jointed appendages; simpler forms lack appendages altogether. Many (e.g., crustaceans, such as shrimp and crabs) have a chitinous exoskeleton that is shed periodically for growth; others (e.g., mollusks, such as oysters and clams) are encased in a heavy, permanent shell that provides protection. Some species do not possess a complete digestive tract, excretory organs, or circulatory or respiratory systems or organs, and most do not have true segmentation (repetition of parts).

Invertebrates eat plants or other animals, generally those smaller and lower on the evolutionary scale than themselves: Fish-eating jellyfish and mollusk-eating starfish are exceptions to this rule. Invertebrates are eaten by a wide variety of fish and other invertebrates, and some are cannibalistic. As a rule, they serve as intermediate or *paratonic* (transport) hosts for parasites. Those that are *sessile* (very slow moving) and have large shells live in close association with many other species of plants and animals.

REFERENCES

Berg, C. J. 1983. *Culture of marine invertebrates: Selected readings.* Stroudsburg, PA: Hutchinson Ross.

Huner, J. V., and E. E. Brown, eds. 1985. *Crustacean and mollusk aquaculture in the United States.* Westport, CT: AVI Publishing Co.

McVey, J. P., ed. 1983. *CRC handbook for mariculture. Vol 1: Crustacean aquaculture.* Boca Raton, FL: CRC Press.

Arthropods/Crustaceans

A strong exoskeleton composed of chitin covers the exterior of crustaceans. In addition to providing protection for the animals, the exoskeleton provides attachment for muscles and forms levers for body parts. Growth necessitates *molting* (shedding), with a new, larger exoskeleton forming inside the old.

Crustaceans serve as host to many species of microorganism, including viruses and fungi, and a wide variety of parasites and symbionts. Their exoskeletons serve as suitable substrate for settlement by many epibiotics. In an effort to reduce the burden of heavy fouling, and perhaps to prevent interference with chemo- and mechanoreception by their *antennal flagella* (sites of taste and tactile reception), crustaceans clean and preen themselves. Such grooming may also reduce the numbers of symbionts on the exoskeleton, thereby preventing loss of mobility resulting from the settlement of symbionts in a joint. Molting too rids the host of unwanted symbionts; but since older crustaceans molt less often than young ones, symbionts may become established for relatively long intermolt periods.

Crustaceans eat fresh and decaying fish or meat; vegetation, including roots, shoots, and leaves of common seaweeds; clams, oysters, and worms; and other crustaceans. They are, in turn, eaten by predatory fishes, such as groupers, jewfish, snappers, bottom fish, and sharks, as well as by other crustaceans (some species are cannibalistic).

When most people think of *soft-shelled* (recently molted) crustaceans, they think only of soft-shelled crabs. However, other crustaceans, such as lobsters, shrimp, and crayfish, can bring higher market prices with a soft shell than with a hard one. Soft-shell markets are already established in the United Kingdom and Southeast Asia, but still require promotion in the United States; those in countries like Japan can be dramatically increased with virtually no new technology. Product image is an important factor: To reach a broader market, soft-shelled crustaceans must be seen as belonging not only in a gourmet seafood restaurant but also on supermarket shelves.

REFERENCES

Homziak, J. 1989. Producing soft crawfish: Is it for you? *Aquaculture Magazine* 15(1):26–32.

Oesterling, M. J., and A. J. Provenzano. 1985. Other crustacean species. In *Crustacean and mollusk aquaculture in the United States*, ed. J. V. Huner and E. E. Brown, pp. 203–34. Westport, CT: AVI Publishing Co.

Wear, R. G. 1990. Soft-shelled crustaceans; new horizons for aquaculture. *World Aquaculture* 21(1):36–8.

Crabs

Crabs vary greatly in size, from a few inches to over 5 ft across the outstretched legs. True crabs possess a small, symmetrical abdomen that is bent and tucked under the depressed *cephalothorax* (carapace). They have five pairs of legs (as do lobsters); the first is modified into *chelae* (pincers), and is usually larger than the others. The most posterior legs are either modified to aid in swimming or are held under the body. Unlike some other crustaceans, such as lobsters and shrimp, crabs lack abdomenal legs (*uropods*). If a leg is injured or seized by an enemy, the crab can drop it at a "breakage" point (where healing and regrowth occurs) and grow another (*regeneration*).

There are several aspects of the biology and market value of crabs that make them difficult to rear for profit. The cost of holding and feeding is generally high. Moreover, many crab species demonstrate aggressive cannibalistic behavior, which may be more pronounced when insufficient food or territory is avail-

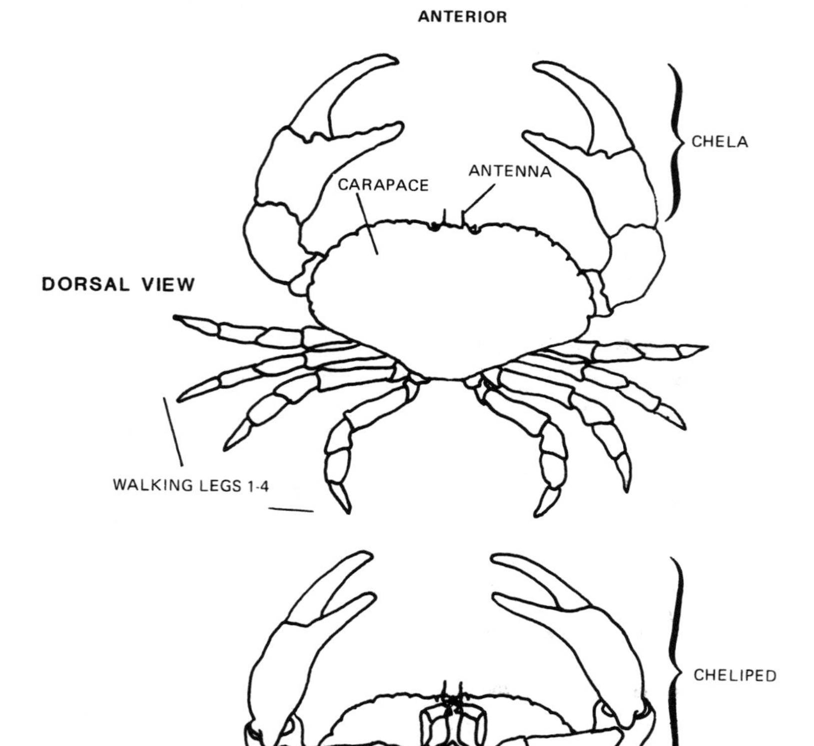

Crab descriptive terminology.

able. Processing of some crabs (e.g., the blue crab) is rather labor intensive: The meat has to be picked from the shell for market, and considerable care must be taken to prevent shell splinters from mixing with it. Market prices may be rather low due to competition both from local fisheries and from imported crabs and crab products.

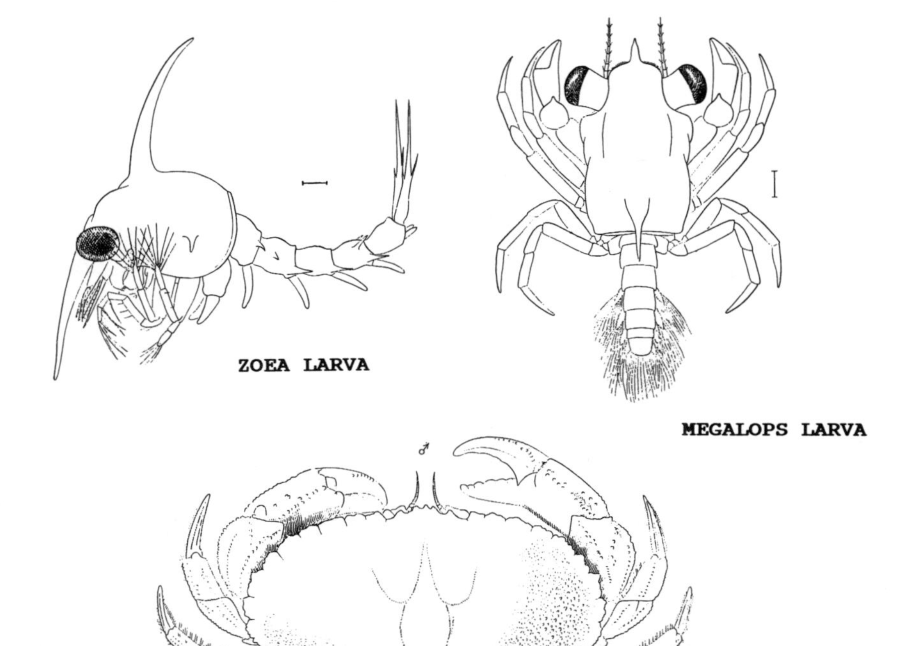

Crab life stages.

REFERENCE

Oesterling, M. J., and A. J. Provenzano. 1985. Other crustacean species. In *Crustacean and mollusk aquaculture in the United States,* ed. J. V. Huner and E. E. Brown, pp. 203–34. Westport, CT: AVI Publishing Co.

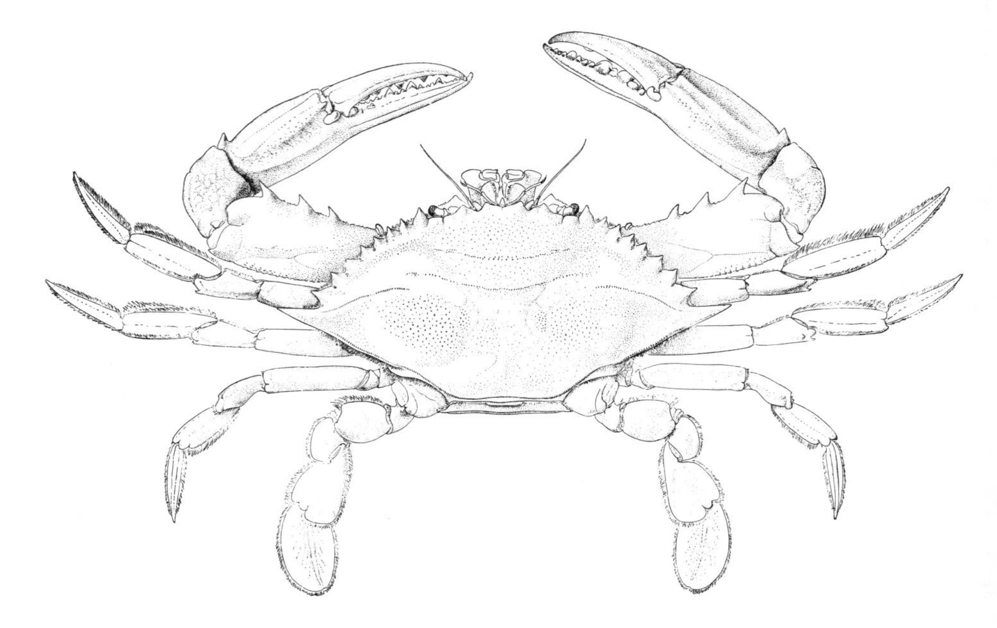

SCIENTIFIC AND COMMON NAMES: ***Callinectes sapidus,* Blue crab**

DESCRIPTION AND DISTINCTIVE CHARACTERISTICS: Long, slender walking legs, with the last pair modified into swimming paddles. The carapace is about 2.5 times as broad as it is long. There are eight distinct teeth on the carapace margin between the eye and the strong lateral spine located at the widest point of the carapace. Color is generally grayish to blue-green, with varying shades of red on the tips of the lateral spines.

HABITAT RANGE: U.S. East Coast from Massachusetts southward around the northern Gulf of Mexico to Texas; transplanted to Europe. Salinities are low in the upper reaches of estuaries and high offshore.

REPRODUCTION AND DEVELOPMENT: Mating occurs in low-salinity waters; spawning takes place offshore in high salinities. Females produce about two million eggs in a "sponge" attached to the abdomen. Hatching occurs in 11–14 days. Larvae pass through two major stages: the *zoea,* lasting roughly 30–50 days, and *megalops,* lasting 6–20 days. In these stages they drift for about 90 days until they assume crablike form and take up a bottom-dwelling existence.

AGE AND GROWTH: The adult stage is reached in 12–18 months from hatching, by the end of which the carapace may be 6 in. wide. About 2.5 years are required to reach the maximum size of roughly 7 in. across the carapace.

FOOD AND FEEDING: Mostly fresh and decaying fish, with some vegetation.

PARASITES AND DISEASE: Viruses, bacteria, fungi, protozoans (several spp.), trematodes, nemerteans, crustaceans (the sacculinid barnacle causes castration), and annelids.

PREDATORS AND COMPETITORS: Competition with other crabs; eaten by various crab species (including its own), octopuses, and fish, as well as by some land animals that search tidal flats for food (e.g., raccoons).

AQUACULTURAL POTENTIAL: Potential for profit not high. The relatively low market price, high cost of holding and feeding, and cannibalism are disadvantages. However, considerable interest has developed recently in the time-honored (since the 1800s) practice of holding hard-shelled crabs that are nearly ready to molt for the soft-shelled market. This short-term aquaculture is popular because the meat does not have to be picked from the shell: The entire body of a soft-shelled crab can be eaten once cooked. Soft-shelled crabs must be removed from salt water within a few hours after shedding of the old exoskeleton. They are considered a delicacy and bring higher prices than hard-shelled crabs.

REGIONS WHERE FARMED AND/OR RESEARCHED: Virginia Institute of Marine Science, Gloucester Point; Mississippi, Delaware, North Carolina.

REFERENCE

Oesterling, M. J., and A. J. Provenzano. 1985. Other crustacean species. In *Crustacean and mollusk aquaculture in the United States*, ed. J. V. Huner and E. E. Brown, pp. 203–34. Westport, CT: AVI Publishing Co.

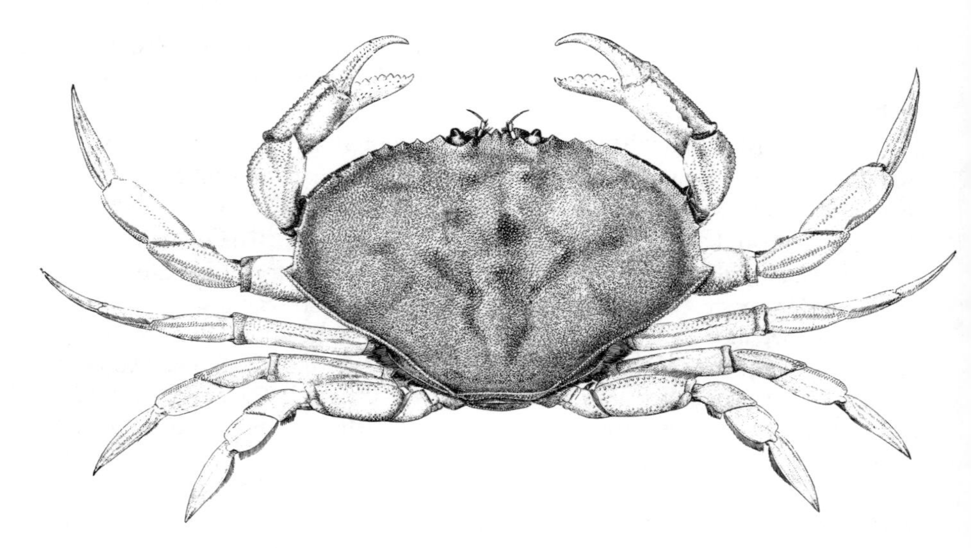

SCIENTIFIC AND COMMON NAMES: ***Cancer magister*, Dungeness crab**

DESCRIPTION AND DISTINCTIVE CHARACTERISTICS: Decapod crustacean with a hard outer shell (*exoskeleton*) that must be shed periodically to allow for growth. The upper side of the carapace is reddish-brown and the underside is whitish.

HABITAT RANGE: U.S. West Coast from the Alaskan Peninsula south to Mexico. Entire life is spent in the sea.

REPRODUCTION AND DEVELOPMENT: Dungeness crabs become sexually mature during their second year. Mating takes place during the summer (usually April–September). As many as three to five million eggs may be produced by one female during a season. They are carried by the female until hatching, after which they live as pelagic larvae near the surface of the water before settling to the bottom.

AGE AND GROWTH: Long larval life. Growth rate is slow: Perhaps 3 or 4 years may be required before crabs are ready for market at a carapace width of about 10 in. Believed to live 8–10 years.

FOOD AND FEEDING: Mollusks, including oysters and cockles, and small fish serve as food; species is cannibalistic.

PARASITES AND DISEASE: Fungi (infect eggs), protozoans, trematodes, nemerteans (on egg masses), and annelids.

PREDATORS AND COMPETITORS: A wide variety of animals prey on adult Dungeness crabs, including octopuses, sharks, sole, lingcod, cabezon, and salmon. Worms feed on the egg mass attached to females. Cannibalism is a major problem during larval and later crab stages.

AQUACULTURAL POTENTIAL: Suggested for restocking dwindling natural supplies and for private ventures. The greatest expense in the rearing of Dungeness crabs is food. Slow growth rate to market size is a disadvantage. There are no reliable economic studies dealing with the profitability of large-scale rearing of this crab. Low market prices and wide price fluctuations lessen the interest in commercial crab culture.

REGIONS WHERE FARMED AND/OR RESEARCHED: Humboldt State University (Arcata, California).

REFERENCES

Cheney, D. P., and T. E. Mumford. 1986. *Shellfish & seaweed harvests of Puget Sound.* Seattle: Washington Sea Grant Program.

Oesterling, M. J., and A. J. Provenzano. 1985. Other crustacean species. In *Crustacean and mollusk aquaculture in the United States,* ed. J. V. Huner and E. E. Brown, pp. 203–34. Westport, CT: AVI Publishing Co.

Wild, P. W., and R. N. Tasto, eds. 1983. *Life history, environment, and mariculture studies of the Dungeness crab,* Cancer magister, *with emphasis on the central California fishery resource.* Fish Bulletin no. 172. Long Beach: California Department of Fish and Game.

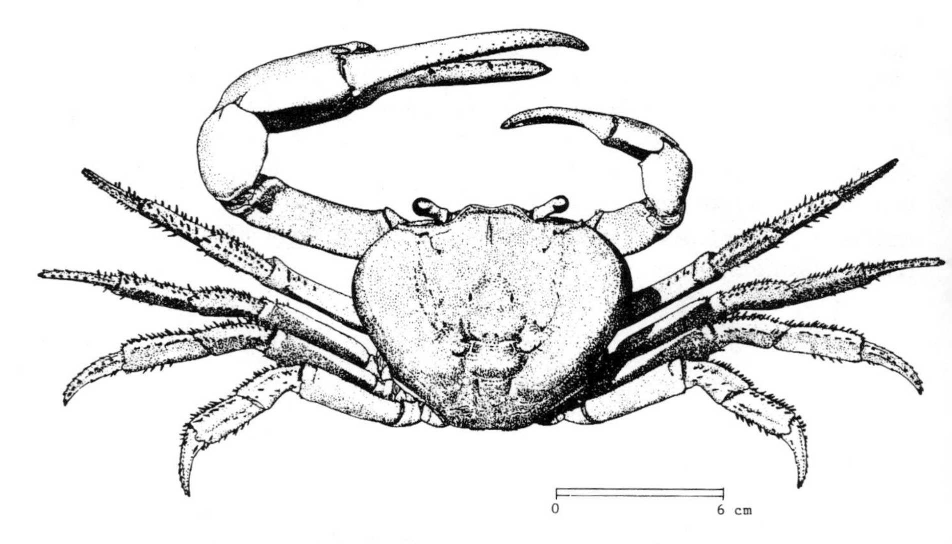

SCIENTIFIC AND COMMON NAMES: ***Cardisoma guanhumi*, Great land crab**
Also called blue land crab.

DESCRIPTION AND DISTINCTIVE CHARACTERISTICS: The great land crab is a large crab that may reach over 17 oz and measure about 6 in. across the back when fully grown. Its claws may be about 6 in. long, with one longer than the other; they are capable of inflicting a sharp pinch. Adult coloration is lavender blue but subject to frequent changes.

HABITAT RANGE: South Florida and throughout the Caribbean and Bahamas. Great land crabs live in burrows that may be several feet deep and usually are dug into the water table. They can tolerate salinities from 0 to full-strength sea water, and can stay submerged for long periods.

REPRODUCTION AND DEVELOPMENT: Eggs are carried beneath the female's body and hatch in shallow inshore waters, from which the tiny larvae are swept to sea. Development to settling crab stages takes over a month, which allows ocean currents to carry them far before they reach another shore.

AGE AND GROWTH: Little is known about their longevity.

FOOD AND FEEDING: Land crabs, though omnivorous, are mainly vegetarians, preferring tender leaves, fruits, berries, and flowers. They mostly eat native plants and agricultural crops, and may be a pest to the home gardener and truck farmer. They also eat large insects, such as beetles. Near human settlements, they are believed to eat most types of human organic refuse.

PARASITES AND DISEASE: Shell disease syndrome (erosion and pitting of the exoskeleton caused by chitin-destroying microorganisms); bacteria and fungi of several genera are implicated in the disease.

PREDATORS AND COMPETITORS: The great land crab is cannibalistic, and is probably also eaten by raccoons and rats.

AQUACULTURAL POTENTIAL: In some Caribbean countries, land crabs are captured at night when they leave their burrows, then held in fenced areas for later sale. They are fed bread, corn, lettuce, watermelon rinds, and cabbage, and are said to be "cleaned" and "fattened" and to have a sweeter taste. Any market that might develop would be limited in geographic range to the tropics and subtropics: Land crabs are commercially important in Puerto Rico and the Bahamas, but their unfamiliarity to consumers elsewhere would make marketing difficult. The hepatopancreas in the carapace (called the fat) is frequently more highly prized as a cooking ingredient than is the muscle tissue.

REGIONS WHERE FARMED AND/OR RESEARCHED: South Florida, the Bahamas, and Puerto Rico.

REFERENCES

Feliciano, C. 1962. *Notes on the biology and economic importance of the land crab,* Cardisoma guanhumi *Latreille, in Puerto Rico.* Institute of Marine Biology, University of Puerto Rico, Contribution no.36. Mayaguez: University of Puerto Rico.

Gifford, C. A. 1962. Some observations on the general biology of the land crab, *Cardisoma guanhumi* (Latreille), in south Florida. *Biological Bulletin* 123:207–23.

SCIENTIFIC AND COMMON NAMES: ***Menippe mercenaria,* Stone crab**

DESCRIPTION AND DISTINCTIVE CHARACTERISTICS: A large crustacean (up to 4.5 in. across the carapace) with powerful claws. The body is dark reddish-brown, spotted with gray, and about two-thirds as long as it is wide. The walking legs have red and yellow bands; claw ends are black.

HABITAT RANGE: North Carolina through the Gulf of Mexico to Mexico; in shallow, nearshore water (10–60 ft deep).

REPRODUCTION AND DEVELOPMENT: A mature female can produce as many as six million eggs in a spring–summer spawning season. She carries the egg masses on her abdomen like an apron for about two weeks, then migrates offshore to saltier water for hatching. Young are free-swimming when they hatch. After numerous molts over the course of a month, they assume true crab form and take up a benthic existence.

AGE AND GROWTH: About three years are required for the crab to reach adult size, with claws large enough for the commercial market; about one year, in nature, is necessary to regenerate claws to legal size.

FOOD AND FEEDING: Varied, including polychaetes, sea grasses, oysters, oyster drills, barnacles, conches, flatworms, blue crabs, hermit crabs, jellyfish, carrion, and other stone crabs (species is cannibalistic).

PARASITES AND DISEASE: Shell disease (bacteria and fungi), protozoans (*Nematopsis*).

PREDATORS AND COMPETITORS: Horse conch, sea turtles, octopuses, fish (cobia, grouper).

AQUACULTURAL POTENTIAL: Meat from stone crab claws is a delicacy: When boiled and dipped in lemon butter, it is considered one of the finest of seafoods. However, artificial cultivation does not appear practical at this time. Stone crabs are territorial and become cannibalistic when crowded together. Preventing cannibalism necessitates large rearing ponds or some form of isolation of individual crabs. Reproduction in captivity is difficult to achieve. The long regeneration time of removed claws is a disadvantage in farming.

REGIONS WHERE FARMED AND/OR RESEARCHED: University of Miami, RSMAS.

REFERENCES

Williams, A. B. 1965. Marine decapod crustaceans of the Carolinas. *Fishery Bulletin* 65:1–298.

Yang, W. T. 1973. Notes on the successful reproduction of stone crabs, *Menippe mercenaria* (Say) reared from eggs. *Proceedings of the 3rd Annual Workshop, World Mariculture Society, 1972*, pp. 183–4.

Yang, W. T., and G. E. Krantz. 1976. *"Intensive" culture of the stone crab,* Menippe mercenaria. University of Miami Sea Grant tech. bull. no. 35. Coral Gables, FL: Sea Grant Program, University of Miami.

SCIENTIFIC AND COMMON NAMES: ***Mithrax spinosissimus,* West Indian king crab**
Also called the spiny crab, Caribbean king crab, or red spider crab.

DESCRIPTION AND DISTINCTIVE CHARACTERISTICS: May reach 18 lb, with a shell length of 7 in. and a width of 2 ft across outstretched claws. Powerful *chelipeds* (claw-bearing appendages) in adult males; female chelipeds are smaller and weaker. Carapace has tubercles or spinules and is usually covered with encrusting organisms. Walking legs are deep red; carapace carmine or red with yellowish tints. Chelipeds may be rose red with yellow tips.

HABITAT RANGE: Tropical western Atlantic, from coastline out to depths of about 100 fathoms; perhaps as far north as North Carolina and south through the Florida Keys and West Indies to Guadeloupe. Usually found around rocks.

REPRODUCTION AND DEVELOPMENT: Spawning occurs August–November in the Florida Keys. Females retain eggs; larval development is abbreviated.

AGE AND GROWTH: In cages may reach 1.25 lb in 250 days and 4.5 lb in just over a year.

FOOD AND FEEDING: Omnivorous, but mostly eats as many as 60 different species of blue-green, brown, and red algae.

PARASITES AND DISEASE: No record.

PREDATORS AND COMPETITORS: Other crabs and fish prey on this species.

AQUACULTURAL POTENTIAL: Described by some observers as nonaggressive, but as very aggressive and cannibalistic by others; the latter behavior, when predominant, is a serious disadvantage to rearing this crab to market sizes in cages. On the positive side, the high market value (due to short supplies of other long-legged crabs, such as the Pacific king crab), short larval life, ease of spawning in captivity, diet of mostly inexpensive algae, and (generally) noncannibalistic behavior make the giant West Indian king crab a good candidate. Caged crabs feeding on "cultured algal turfs" did not suggest that commercial ventures would be successful, however: Some rearing trials showed this crab to be highly cannibalistic when lacking animal protein in the diet. Furthermore, the algal rearing techniques were labor intensive.

REGIONS WHERE FARMED AND/OR RESEARCHED: Harbor Branch Oceanographic Institution (Fort Pierce, Florida); Los Roques, Venezuela; Turks and Caicos Islands, Dominican Republic, and Antigua.

REFERENCES

Abey, W. H. 1985. *Summary of Caribbean king crab* (Mithrax spinosissimus) *mariculture development.* Washington, DC: Marine Systems Laboratory, Smithsonian Institution (unpublished).

Alston, D. E. 1991. Culture of crustaceans in the Caribbean. *World Aquaculture* 22(1):64–8.

Brownell, W. N., A. J. Provenzano, and M. Martinez. 1977. Culture of the West Indian spider crab (*Mithrax spinosissimus*) at Los Roques, Venezuela. *Proceedings of the World Mariculture Society* 8:157–68.

Williams, A. B. 1984. *Shrimps, lobsters, and crabs of the Atlantic coast of the eastern United States, Maine to Florida.* Washington, DC: Smithsonian Institution Press.

Crayfish

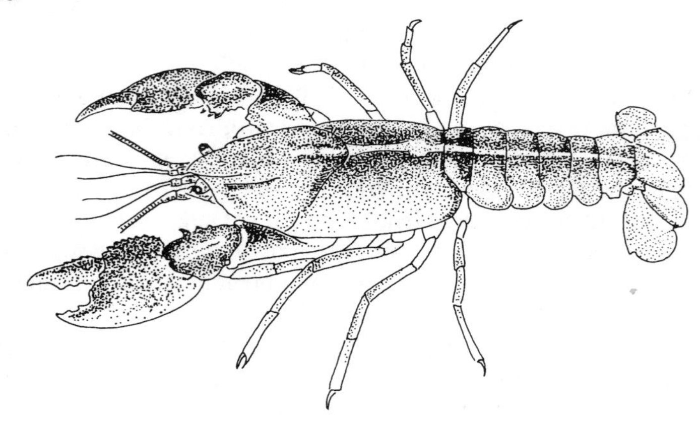

Red swamp crawfish.

SCIENTIFIC AND COMMON NAMES: There are four genera and 29 species of crayfish (also called crawfish, crawdads, mudbugs, Dixie lobsters, and freshwater lobsters). The terms "crawfish" and "crayfish" are interchangeable, but whereas "crayfish" is preferred in the scientific community, "crawfish" is preferred commercially (Avault and Huner 1985). Only two large species, the red swamp crawfish (*Procambarus clarkii*) and the white river crawfish (*P. blandingii acutus*) are taken for commercial markets; smaller species are used as bait by sportfishers.

DESCRIPTION AND DISTINCTIVE CHARACTERISTICS: These small crustaceans are related to lobsters and show a close resemblance. The carapace of adults is dark red to nearly black, with a wedge-shaped black stripe on the abdomen.

HABITAT RANGE: Westward to Texas and southeastern New Mexico temperate climates; other species are found west of the Rocky Mountains, and still others live in many other countries worldwide. They are freshwater dwellers, burrowing into stream banks and swamp and pond bottoms. They can survive in stagnant, turbid, low-oxygen waters.

REPRODUCTION AND DEVELOPMENT: Adults inhabit open waters in winter and spring. About July 1, females dig burrows 24–36 in. deep in dry areas. When fields are still flooded, they burrow in levees, high spots in fields, or migrate to adjacent fields. Egg laying begins about September 1. The eggs are fertilized by a sperm receptacle that the male placed on the female in May, and deposited in balls on the swimming legs. Red swamp crawfish produce 200–400 young in 18–29 days. September and October appear to be the peak hatching times, but adults have been observed with young every month of the year. If water is available, crayfish will come out of their holes about October 1 to rear and scatter their young.

AGE AND GROWTH: About 210 days are required for late-summer- or fall-hatched red swamp crawfish to reach their minimum marketing size of ~0.5–0.7 oz. Molting of the exoskeleton is required for growth.

FOOD AND FEEDING: Crayfish feed best at temperatures of about 65–80 °F; little feeding occurs below 45° or above 90°. Crayfish will eat animal matter (insect larvae, etc.) in small amounts, but are not active predators; dead and living vegetation make up the bulk of their diet. They can be cannibalistic: Females will eat a portion of their brood when confined in small areas without food.

PARASITES AND DISEASE: The fungus *Aphanomyces astaci,* which occurs in American crayfishes, is believed to have caused a crayfish plague in Europe after the introduction of American species. Fungus (apparently like *Saprolegnia*) also occurs on eggs. Filamentous bacteria clog gills. Viruses and rickettsia and trematodes are also parasitic. Numerous organisms associate with crayfish but do not appear to be parasites, including algae, rotifers, leeches, nematodes, insects, isopods, ostracods, copepods, turbellaria, and barnacles.

PREDATORS AND COMPETITORS: Fish are the worst predators of crayfish, but raccoons, ibises, herons, boat-tailed grackles, sea gulls, legless salamanders, predacious water beetles, and dragonflies are all known to eat crawfish.

AQUACULTURAL POTENTIAL: Some farmers rear red swamp crawfish as their main crop and manage impoundments either for this purpose alone or with the possibility of developing crops of bullfrogs and turtles. Others raise crayfish in rice fields, or in rotation with rice: Crayfish are a winter crop and rice a summer crop. Annual yields from a good pond vary from 400 to 1,000 lb/acre; experimental production with supplemental feeding yielded 1,200 lb/acre. Individuals harvested in December to early February are large, averaging 10–11/lb; average size then decreases as young of the year enter the catch. The market for soft-shelled (recently molted) crayfish is stronger than that for normal hard-shell crayfish; moreover, it is predicted to increase rapidly. Further market promotion may be needed to shake crayfish's image as a strictly southern seafood. In addition to crayfish species raised for human food, several are produced in many other states for the bait market, serving as excellent bait for the sunfishes, yellow perch, bass, and even catfish.

REGIONS WHERE FARMED AND/OR RESEARCHED: The state of Louisiana produces about 99 percent of the hundred million pounds of crayfish sold in the United States for human food. Small amounts are also produced in Arkansas, Mississippi, Texas, and South Carolina.

REFERENCES

Avault, J. W., and J. V. Huner. 1985. Crawfish culture in the United States. In *Crustacean and mollusk aquaculture in the United States*, ed. J. V. Huner and E. E. Brown, pp. 1–61. Westport, CT: AVI Publishing Co.

De la Bretonne, L. 1977. A review of crawfish culture in Louisiana. *Proceedings of the World Mariculture Society* 8:265–9.

Goldman, C. R., ed. 1983. *Freshwater crayfish V*. Westport, CT: AVI Publishing Co.

Homziak, J. 1989. Producing soft crawfish: Is it for you? *Aquaculture Magazine* 15(1):26–32.

Huner, J. V. 1976. Raising crawfish for food and fish bait: A new polyculture crop with fish. *Fisheries* 1(2):7–9.

Huner, J. V. 1988. Crayfish culture in Europe. *Aquaculture Magazine* 14(2):48–52.

Huner, J. V. 1991. Aquaculture of freshwater crayfish. In *Production of aquatic animals; crustaceans, molluscs, amphibians and reptiles*, ed. C. E. Nash, pp. 45–66. Amsterdam: Elsevier.

Huner, J. V., and J. E. Barr. 1984. *Red swamp crawfish: Biology and exploitation*, rev. ed. Baton Rouge: Louisiana Sea Grant College Program.

Romaire, R. B., ed. 1989. Proceedings of the special symposium: Crawfish industry status and trends, New Orleans, 1988. *Journal of Shellfish Research* 8(1):255–313.

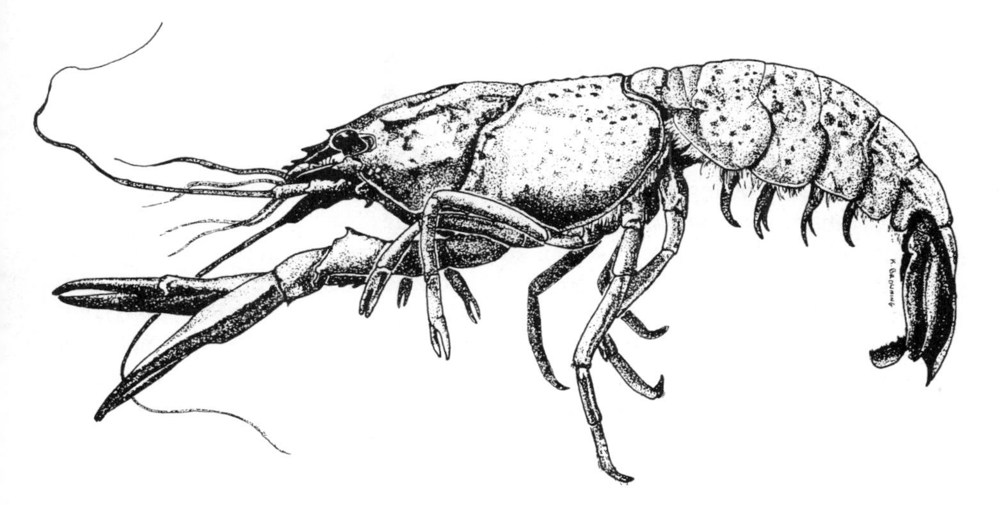

SCIENTIFIC AND COMMON NAMES: ***Cherax tenuimanus*, Marron**
Of the hundred or so species of crayfish in Australia, only three species in the genus *Cherax* seem to have aquaculture potential. *C. tenuimanus* has received the greatest research and commercial attention.

DESCRIPTION AND DISTINCTIVE CHARACTERISTICS: Hailed as one of the biggest freshwater crayfish in the world. Has five ridges or keels on the back of the head.

HABITAT RANGE: Natural habitat in freshwater rivers in extreme southwestern Australia, from Geraldton south and east to Esperance; there they find cool water suitable to their well-being. Transplanted to many different countries.

REPRODUCTION AND DEVELOPMENT: Spawn in spring or early summer. Not highly fecund: 450–900 fertilized eggs carried for about 12–16 weeks by females, depending on size of the mother. Offspring are similar to adults, except in coloration and size, when they leave the mother. Some marrons may spawn as early as 2 years of age, but most do not spawn until they are 3.

AGE AND GROWTH: Exceptional specimens are known to reach a head–tail length of 15.5 in. and weigh up to 6 lb. Wild marron generally average about 7–9 oz; in ponds marron usually reach a size of 2 oz in about a year.

FOOD AND FEEDING: Eat dead plants, animals, and other detritus.

PARASITES AND DISEASE: Mass mortalities due to pathogens unknown. May be parasitized by the microsporean *Thelohania*. Fungus diseases found in warm-water crayfish have not been reported in marrons.

PREDATORS AND COMPETITORS: Cannibalism by females on newly molted young. Aggressive about shelter. Marron in clear water are preyed upon by birds.

AQUACULTURAL POTENTIAL: High meat per animal, about 60 percent edible meat (40% from the abdomen, ~16% from the claws). Relatively free from parasites and disease. Survival is high when held briefly out of water in a damp, cool atmosphere. Densities up to about 5/m² can be used for small marrons. High market price is a plus. Growth is rather slow: Even after two years the average size in growing ponds may be below Australia's legal size of 4 oz.

The marron was introduced to the United States from Australia in the 1970s; their aquacultural potential here is limited, however, because few U.S. locations have the year-round temperatures of 68–77 °F they require for optimum growth. Recent research shows that the red claw (*Cherax quadricarinatus*), native to a large area in northwestern Australia, may have greater U.S. potential due to its temperature range and rapid growth. Its simple life cycle and tolerance to rather harsh pond conditions are added advantages. Still, much more biological and market research is required before the red claw can be transplanted to the U.S. Southeast; even then, resistance can be expected, as with any other exotic species.

REGIONS WHERE FARMED AND/OR RESEARCHED: Australia, New Zealand, Europe, Africa, Asia, Dominica (Eastern Caribbean), and the United States.

REFERENCES

Aiken, D. 1988. Marron farming; a real industry or just great promotion? *World Aquaculture* 19(4):14–17.

Crook, G. 1985. *Marron and marron farming*. Fisheries Information Publication no. 4. Perth: Extension and Publicity Office, Fisheries Dept., Western Australia.

Morrissy, N. M. 1976a. *Aquaculture of marron,* Cherax tenuimanus *(Smith). Site selection and the potential of marron for aquaculture.* Fisheries Research Bulletin no. 17, pt. 1. Perth: Fisheries Dept., Western Australia.

Morrissy, N. M. 1976b. *Aquaculture of marron,* Cherax tenuimanus *(Smith). Breeding and early rearing*. Fisheries Research Bulletin no. 17, pt. 2. Perth: Fisheries Dept., Western Australia.

Rouse, D. B., C. M. Austin, and P. B. Medley. 1991. Progress toward profits? Information on the Australian crayfish. *Aquaculture Magazine* 17(3):46–56.

Rubino, M., N. Alon, C. Wilson, D. Rouse, and J. Armstrong. 1990. Marron aquaculture research in the United States and the Caribbean. *Aquaculture Magazine* 16(3): 27–44.

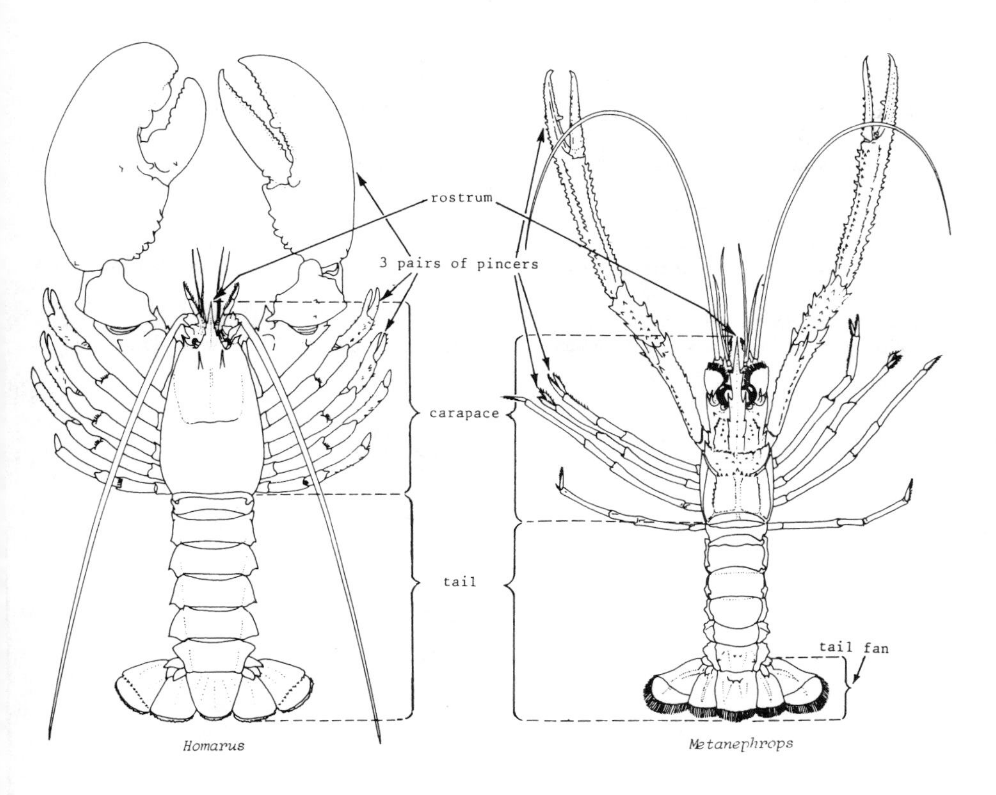

Lobster descriptive terminology.

Lobsters

These crustaceans range in length from a few inches to more than a yard. The body is elongate, tubular, or flattened and has a prominent "tail" or abdomen made up of six movable segments that terminate in a fan (*telson*). The anterior end (*carapace*) is rigid and about the same length as the abdomen. Some species are covered with spines, and some have rather smooth exoskeletons. Lobsters have stalked, movable eyes, and most have long, strong antennae. There are five pairs of walking legs; in some groups the first pair are modified as pincers (*chelae*). Legs on the abdomen are short and biramous (*pleopods*). Lobsters occur off both coasts of the United States.

REFERENCES

Cobb, J. S., and B. F. Phillips, eds. 1980. *The biology and management of lobsters,* 2 vols. New York: Academic Press.

D'Abramo, L. R., and D. E. Conklin. 1985. Lobster aquaculture. In *Crustacean and mollusk aquaculture in the United States,* ed. J. V. Huner and E. E. Brown, pp. 159–201. Westport, CT: AVI Publishing Co.

Nash, C. E. The production of marine lobsters. 1991. In *Production of aquatic animals; crustaceans, molluscs, amphibians and reptiles,* ed. C. E. Nash, pp. 79–85. Amsterdam: Elsevier.

Taylor, H. 1984. *The lobster: Its life cycle,* rev. ed. New York: Pisces Books.

Waddy, S. L. 1988. Farming the homarid lobsters: State of the art. *World Aquaculture* 19(4): 63–71.

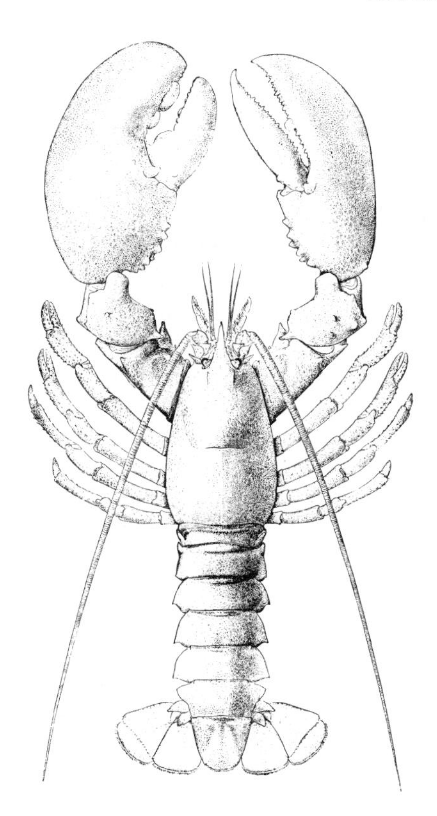

SCIENTIFIC AND COMMON NAMES: ***Homarus americanus,***
Also called Maine lobster or northern lobster. **American lobster**

DESCRIPTION AND DISTINCTIVE CHARACTERISTICS: Decapod crustacean (10 jointed legs) with two large, well-developed claws with teeth and sharp spines. Dark green coloration with splotches of red, blue, and yellow. Infrequently, wholly red, white, or blue lobsters are found in the wild.

HABITAT RANGE: East Coast of North America, from North Carolina to Labrador.

REPRODUCTION AND DEVELOPMENT: Upon hatching from the eggs, embryos are cemented to the hairs of the swimmerets for the last nine months of embryo development. A female may produce 3,000–100,000 eggs during one spawning. Young lobsters pass through several molts and, after a short drifting period, take up a bottom-dwelling existence.

AGE AND GROWTH: About five years are required for the American lobster to reach a market size of 3.3 in. carapace length. Molting of the exoskeleton is required for growth.

FOOD AND FEEDING: Eat seaweed, clams, snails, scallops, starfish, and fish. Estimated that about 15 lb of food are needed to grow 1 lb of lobster.

PARASITES AND DISEASE: Bacteria (including *Gaffkya homari)*, fungi, protozoans, trematodes, nemerteans, acanthocephalans, nematodes, annelids. Shell disease (breakdown of exoskeleton); molt death syndrome.

PREDATORS AND COMPETITORS: Nemertean worms eat eggs. Young lobsters are eaten by many species of bottom fish. Large juveniles and adults are food for large fish (e.g., cod, sharks). Lobsters are cannibalistic when crowded.

AQUACULTURAL POTENTIAL: A popular and desirable seafood, lobsters are hardy and relatively easy to grow; but the long time (5–7 years) required for growth to market size and the high cost of food are significant deterrents to productive farming of this species. Techniques have been developed for reproduction in captivity. Artificial rearing of lobster eggs to larval stages for release in the wild was begun in the United States around the mid-1880s at the dawn of the artificial-propagation frenzy. In more recent times, after about 40 years of scientific research on culturing of all stages of the northern lobster at the Massachusetts Lobster Hatchery and Research Station on Martha's Vineyard, most aspects of the biology of the lobster and culturing techniques have been described. The extended larval development and slow growth to market size are critical problems, as is their cannibalistic behavior. Attempts to increase the growth rate by transplantation to California and to the warm water of Anguilla in the Caribbean were not encouraging. Other problems still needing attention are improved feeds, disease control and treatment, and engineering designs that can provide inexpensive heated seawater, such as power-generating plant effluents. In any case, startup costs will be high and no profits should be expected for about three years.

REGIONS WHERE FARMED AND/OR RESEARCHED: Massachusetts Lobster Hatchery and Research Station; trials at the University of California and the Caribbean island of Anguilla; Maine, Rhode Island; British Columbia, New Brunswick.

REFERENCES

Cobb, J. S., and B. F. Phillips, eds. 1980. *The biology and management of lobsters,* 2 vols. New York: Academic Press.

D'Abramo, L. R., and D. E. Conklin. 1985. Lobster aquaculture. In *Crustacean and mollusk aquaculture in the United States,* ed. J. V. Huner and E. E. Brown, pp. 159–201. Westport, CT: AVI Publishing Co.

Taylor, H. 1984. *The lobster: Its life cycle,* rev. ed. New York: Pisces Books.

Waddy, S. L. 1988. Farming the homarid lobsters: State of the art. *World Aquaculture* 19(4): 63–71.

SCIENTIFIC AND COMMON NAMES: ***Homarus gammarus*, European lobster**
Formerly called *H. vulgaris.*

DESCRIPTION AND DISTINCTIVE CHARACTERISTICS: The differences between the American and European lobster are minor. The most prominent feature of both lobsters is the strong large claws on the first pair of legs. One functions as a feeding claw and the other as a crusher. There are no spines on the various body segments. The color of the European lobster varies from rust to bright blue on the dorsal side, and may be yellowish on the ventral side.

HABITAT RANGE: Mediterranean and Black Sea, the Atlantic coasts of Europe as far north as the Lofoten Islands; south off the western coast of Africa to Agadir. Inhabits rocky bottoms from the coast to depths of about 200 ft.

REPRODUCTION AND DEVELOPMENT: The mother carries the eggs on her swimmerets until they hatch; the pelagic larval stage that follows lasts about two weeks. They have been cultured for about 100 years.

AGE AND GROWTH: After six years, European lobster may reach a total length of 9 in.

FOOD AND FEEDING: Omnivorous; eat benthic invertebrates such as mollusks, starfish, and bottom-dwelling fish.

PARASITES AND DISEASE: Bacterial infestation of eggs and larvae; fungus on eggs and larvae; molt death syndrome.

PREDATORS AND COMPETITORS: Benthic fishes prey on lobsters of all sizes; nemerteans prey on eggs.

AQUACULTURAL POTENTIAL: Although European lobsters have been raised artificially for many years through all life stages, it was not until about 1970 that a substantial effort was mounted to raise them for profit. Many of the deterrents to successful commercial aquaculture that pertain to the American lobster also pertain to the European lobster. In attempts to bypass the difficulties and expense of artificial hatching and rearing of the European lobster, small lobsters were caught and retained until they reached market size. The difficulty inherent in this method is discussed below under spiny lobsters.

REGIONS WHERE FARMED AND/OR RESEARCHED: Scotland and Norway.

REFERENCE

D'Abramo, L. R., and D. E. Conklin. 1985. Lobster aquaculture. In *Crustacean and mollusk aquaculture in the United States,* ed. J. V. Huner and E. E. Brown, pp. 159–201. Westport, CT: AVI Publishing Co.

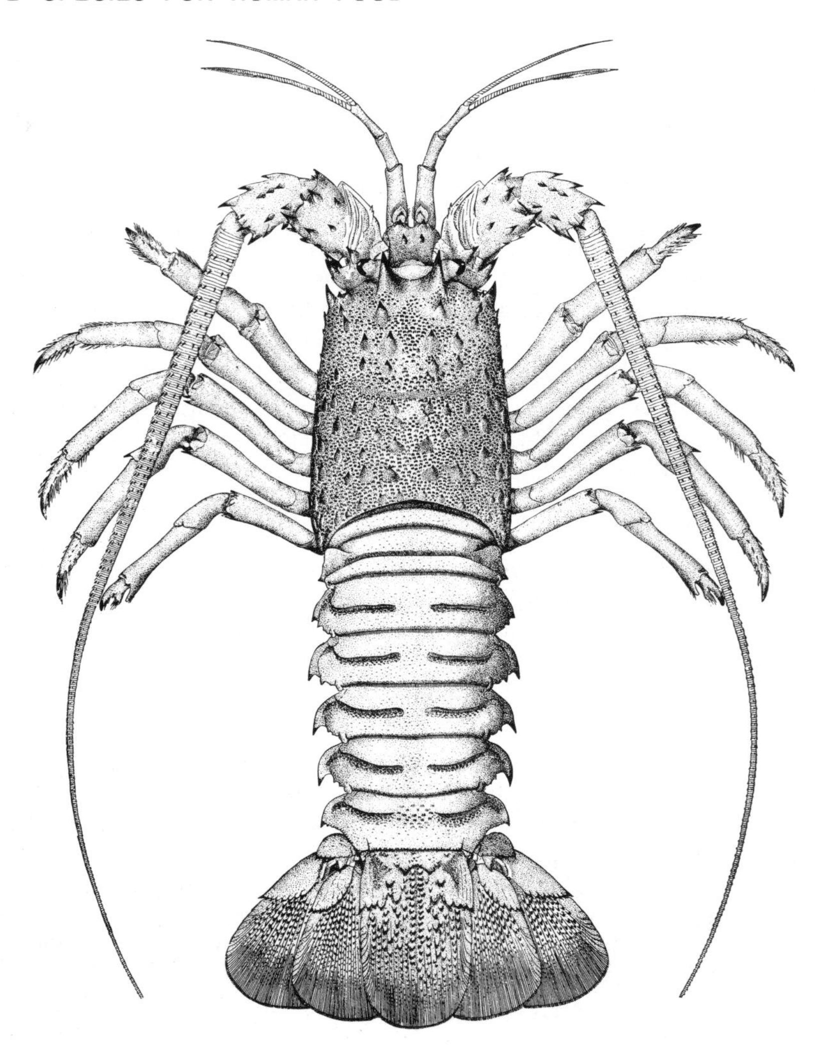

SCIENTIFIC AND COMMON NAMES: ***Panulirus argus*, Spiny lobster** Related: *P. cygnus* (Australia).

DESCRIPTION AND DISTINCTIVE CHARACTERISTICS: Similar in appearance to the American lobster but without large claws or rigid fantail. The spiny lobster has long, whiplike antennae, and its body is covered with forward-projecting spines.

REPRODUCTION AND DEVELOPMENT: Eggs are attached to the abdomen of the female where they are incubated for three to five weeks. The larvae (*phlyllo-*

somes) are cast free in the ocean and drift in the plankton for 6–12 months. Within each larval stage they undergo *instars* (stages between successive molts) generally accompanied by high mortality. Water temperature plays an important role in development.

AGE AND GROWTH: Long larval life, with up to 12 months to settle; thereafter, between 18–36 months may be required to grow from egg to market size. Maximum size of 10 lb.

FOOD AND FEEDING: Eat small mollusks and crustaceans, worms, seaweed, fish, coelenterates, echinoderms, and sponges. The spiny lobster is cannibalistic and a scavenger.

PARASITES AND DISEASE: Very lightly parasitized. Microsporean (protozoan) in tail. Shell disease (bacteria). Stalked barnacle on gills (commensal).

PREDATORS AND COMPETITORS: Pelagic fishes, including tuna (eat larvae), groupers, snappers, other bottom fish, sharks and rays, octopi, and turtles.

AQUACULTURAL POTENTIAL: The cost of raising, handling, and feeding spiny lobsters (if someone could mass-raise them from egg to market size) would be extremely high—so high, in fact, that no profit could be expected. Also problematic is a farming operation based on catching sufficiently large numbers of young in nature when needed: The high cost of collection and of their subsequent maintenance for almost two years could spell economic failure. This has been tried on Antigua in the Caribbean using artificial habitats made of air conditioning filters hung vertically in the water column. Young larvae, approaching settlement and apparently seeking safety, hide between the filters and can be removed and placed in enclosures ashore for growing out. In this experiment, the catches of young in the habitats around Antigua did not suggest that this method could provide sufficent young for a commercial grow-out operation.

REGIONS WHERE FARMED AND/OR RESEARCHED: Harbor Branch Oceanographic Institution (Fort Pierce, Florida), University of Miami, RSMAS; Bermuda; Antigua, British West Indies; Bonaire, Netherlands Antilles, Australia.

REFERENCES

D'Abramo, L. R., and D. E. Conklin. 1985. Lobster aquaculture. In *Crustacean and mollusk aquaculture in the United States*, ed. J. V. Huner and E. E. Brown, pp. 159–201. Westport, CT: AVI Publishing Co.

Miller, D. L. 1983. Shallow-water mariculture of spiny lobster (*Panulirus argus*) in the western Atlantic. In *Proceedings of the First International Conference on Warm Water Aquaculture—Crustacea*, ed. G. L. Rogers, R. Day, and A. Lim, pp. 238–54. Laie, HI: Brigham Young University, Hawaiian Campus.

Tamm, G. R. 1980. Spiny lobster culture: An alternative to natural stock assessment. *Fisheries* 5(4):59–62.

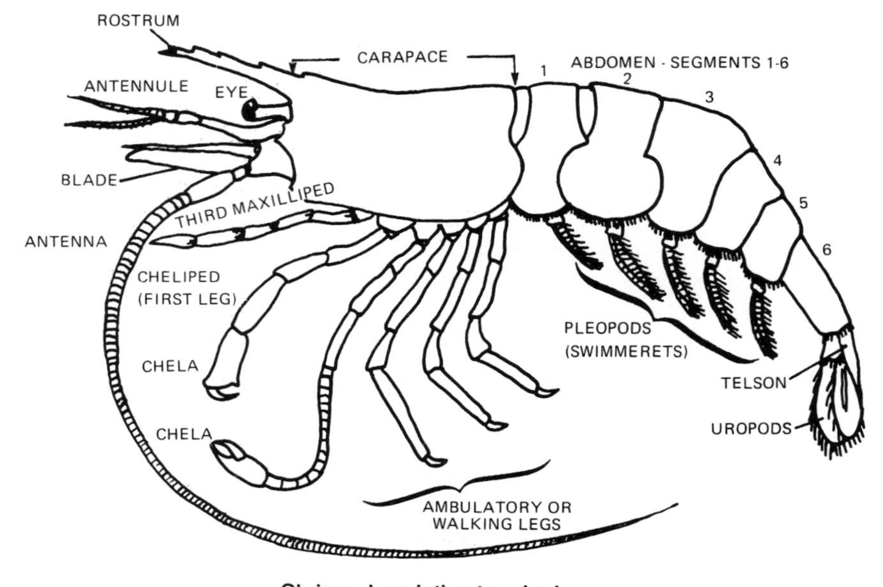

Shrimp descriptive terminology.

Shrimps and Prawns

The terms "shrimp" and "prawn" are generally used interchangeably; however, in some areas, "prawn" is used only to describe large shrimp species. The giant river prawn, a freshwater prawn discussed below, is somewhat larger than its marine relatives and is usually called a prawn.

The shrimp body is nearly always laterally compressed—long and narrow with antennae and stalked eyes. The carapace bears a beaklike structure (*rostrum*) that is usually laterally compressed and toothed. The legs or *pleopods* are usually slender; but in some shrimp, a single leg or pair of legs may be heavy, and some legs end in pincers. The abdominal legs are used for swimming.

Shrimps are favored as aquacultural candidate species owing to several factors: the rapid growth of warm-water shrimp to market size in less than a year; the relatively short larval life (about 3 weeks), limited handling, and processing of harvested shrimp; the extremely high market demand worldwide; and their high price per pound.

Despite the many advantages of farming these species, there is a history of failures in U.S. shrimp farming. Many farming ventures began on a large scale without sufficient background information. Scientific information critical to the advancement of aquaculture was slow in coming because funding and authority struggles developed among federal departments and agencies. State aquacultural research programs pressured the federal government to get involved; when

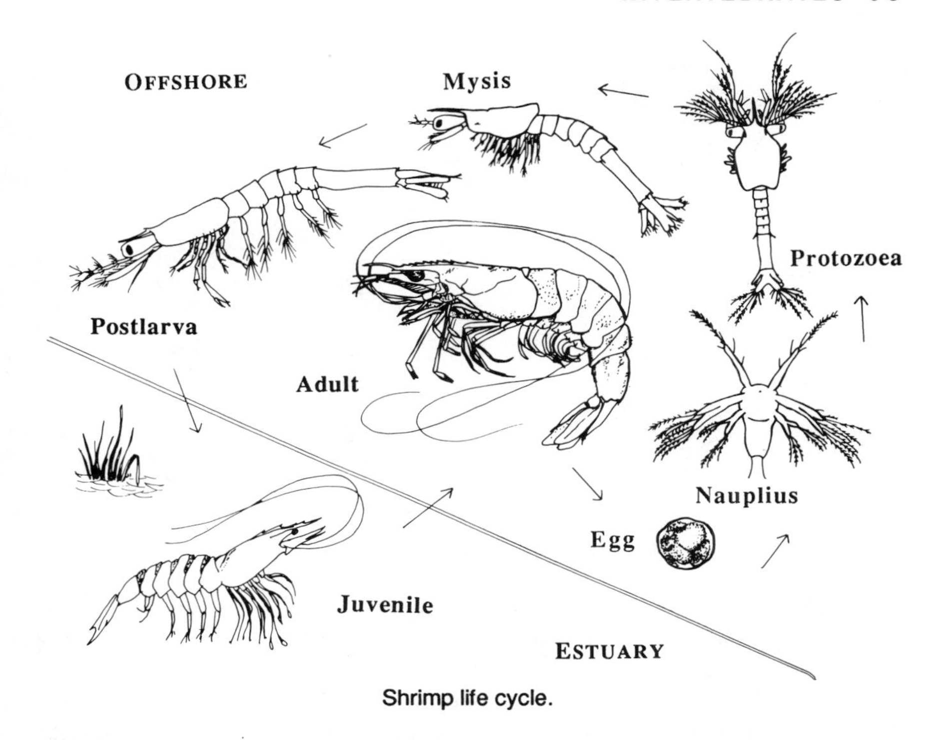

Shrimp life cycle.

the feds dropped the ball, the states again came forward, though frequently with limited budgets. Unlike in some other countries where cooperatives and associations of fish farmers were formed, U.S. shrimp farmers neither shared information nor presented a united front to lobby for governmental assistance to aid the industry.

The high profit required on investments, together with high labor, land, and feed costs, also doomed many trials. However, the U.S. government, private industry research, and pilot plant programs have recently yielded a tremendous store of valuable biological and practical data that can be applied to shrimp culture both in and outside the United States. South Carolina (summer crop only) and Texas managed to farm two million pounds of penaeid shrimp in 1990.

Several local species of warm-water marine shrimp of economic value have been used in U.S. farming trials, including *Penaeus duorarum*, *P. setiferus*, and *P. aztecus*. When it became apparent that exotic shrimp species might do better aquaculturally, at least six of them—*P. vannamei*, *P. stylirostris*, *P. occidentalis*, *P. japonicus*, *P. monodon*, and *P. schmitti*—were brought to the United States to determine their farming suitability. One of these, the giant tiger prawn (*Penaeus monodon*), has considerable appeal to shrimp farmers because it is a very fast-growing species, reaching a length of 13 in. and a weight of 4.5 oz.

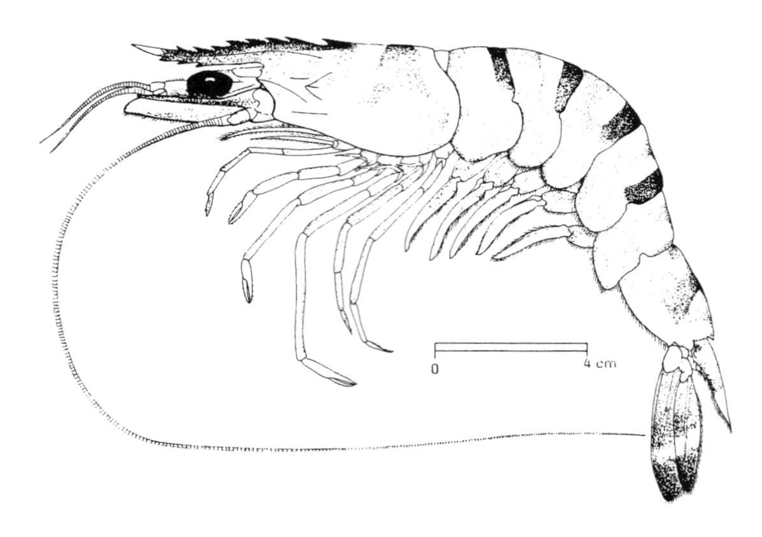

Giant tiger prawn (*Penaeus monodon*).

In the Philippines it is called *sugpo,* but many common names apply throughout its geographic range. Considerable farming technology on this species is available since it is farmed in many areas.

Another U.S.-imported exotic, the kuruma prawn (*Penaeus japonicus*), also has many common names throughout its range. It supports important fisheries in Japan and in the Indo-West Pacific (e.g., the Philippines, Korea, and Taiwan), and extends into the Mediterranean (e.g., France). The female kuruma prawn reaches a total length of about 9 in.; males are usually slightly smaller. Dr. Fuginaga, who alerted the world to the potential of large-scale shrimp farming, used this species in his pioneering research on artificial shrimp rearing.

Permission to bring exotic species into the United States requires a considerable outlay of time and money, and requests are frequently denied. Most permissions have been granted to state, federal, or university agencies for research purposes; few have gone to those engaged in purely commercial ventures.

Warm-water (penaeid) shrimp are usually abundant in areas characterized by an inland brackish marsh connected by passes or outlets to an area of relatively high salinity offshore. Eggs are laid directly in the water offshore; as the young grow they move inshore to "nursery" areas, usually large estuarine bays. Later, as they grow still larger, they move back offshore, and as late juveniles or young adults appear in the deep saline offshore waters.

Members of the genus *Penaeus* are difficult to identify: Morphological characteristics are so minor, especially in the early developmental stages, that positive identification usually requires the knowledge and experience of an expert. Coloration of some species is characteristic and is therefore a useful and quick

Kuruma prawn (*Penaeus japonicus*).

means to identify some species: The kuruma prawn, for example, displays beautiful coloration of the exoskeleton, including brownish stripes or bars on the abdomen (tail section).

Culturing of kuruma prawns in Japan is extremely successful, as is that of giant tiger prawns in the Philippines. Since the biology, life cycle, and rearing techniques of these species are so similar to those of other imports in the genus, only the Pacific white shrimp (*Penaeus vannamei*) is discussed in detail below.

Investigations have been made into the possible use in aquaculture of cold-water (northern) shrimps of the family Pandalidae; but because of their small size and relatively slow growth, most cold-water shrimps do not seem to be viable candidates. One species, the spot prawn (*Pandalus platyceros*), found from San Diego (California) to Alaska, has been cultured from egg through juvenile sizes. Spot prawns are relatively large, easy to maintain in captivity, and accept a wide range of foods. They have been placed in floating net pens in Puget Sound, Washington, and may serve as an additional crop in salmon net pens. Advantages to such salmon–shrimp polyculture is that the shrimp scavenge dead salmon and remove fouling from both the bottom and walls of the net pens.

Some *Penaeus* species are also tolerant to cold waters, growing at temperatures as low as 50 °F, as compared to 68° for others in the genus. The cold-tolerant *P. chinensis, P. japonicus,* and *P. penicillatus* are native to temperate waters at higher latitudes of China, Japan, Taiwan, and Korea; *P. chinensis* is of particular interest since it grows fast and matures in captivity. These species have obvious potential for farming in the United States, where cool fall and spring temperatures shorten the growing season both of local shrimps and of the introduced Pacific white shrimp (*P. vannamei*), a favorite farming species native to warm subtropical waters of South and Central America. There are some drawbacks to using these cold-tolerant species, however: Their protein requirements are sometimes high, raising the cost of feed; also, in some areas, markets would have to be developed. The biggest problem, of course, is the matter of transplanting an exotic.

Plans have been considered for augmenting U.S. wild shrimp stocks by releasing hatchery-reared young shrimp. Similar programs (e.g., for finfish and some invertebrates at small sizes) have a dismal history, being very expensive and generally failing. Although potential advantages are obvious—no long-term care and feeding are required, the growing juveniles find their own food—usually less than 1 percent of hatchery-reared young can be expected to survive in the wild. Thus the numbers of hatchery releases that might be available for harvest is so small as to minimize the chance of a favorable cost–benefit ratio.

Despite these hurdles, Japan holds hope for enhancing wild stocks of *P. japonicus,* and have been experimenting with releasing hatchery-reared shrimp into specially designed, elevated intertidal pools that prevent the entrance of many of the highly predatory fishes that eat shrimp. Releases of 150 million small shrimp per year have been reported, and Japanese calculations suggest substantial profits from these programs. A question has arisen as to who shall be allowed to catch shrimp from stocks that may have benefited from the artificial enhancement: the fishing industry at large, or only those organizations that have built and maintained the intertidal pools.

REFERENCES

Berg, C. J. 1983. *Culture of marine invertebrates: Selected readings.* Stroudsburg, PA: Hutchinson Ross.

Brackishwater Aquaculture Information System. 1988. *Biology and culture of* Penaeus monodon. BRAIS State-of-the-Art Series no. 2. Tigbauan, Iloilo (Philippines): Aquaculture Dept., Southeast Asian Fisheries Development Center.

Chamberlain, G. W., M. G. Haby, and R. J. Miget. 1985. *Texas shrimp farming manual.* Corpus Christi, TX: Texas Agricultural Extension Service, Texas A&M University.

Hanson, J. A., and H. L. Goodwin. 1977. *Shrimp and prawn farming in the Western Hemisphere: State-of-the-art reviews and status assessments.* Stroudsburg, PA: Dowden, Hutchinson & Ross.

Huner, J. V., and E. E. Brown, eds. 1985. *Crustacean and mollusk aquaculture in the United States.* Westport, CT: AVI Publishing Co.

Jung, C. K., and W. G. Co. 1988. *Prawn culture: Scientific and practical approach.* Dagupan City (Philippines): Westpoint Aquaculture Corp.

McVey, J. P., ed. 1983. *CRC handbook for mariculture. Vol 1: Crustacean aquaculture.* Boca Raton, FL: CRC Press.

Main, K. L., and W. Fulks. 1990. *The culture of cold-tolerant shrimp: Proceedings of an Asian–U.S. workshop on shrimp culture, 1989.* Honolulu, HI: Oceanic Institute.

Mock, C. R. 1973. Shrimp culture in Japan. *Marine Fisheries Review* 35(3–4):71–4.

New, M., and S. Singholka. 1985. *Freshwater prawn farming: A manual for the culture of* Macrobrachium rosenbergii, rev. ed. FAO Fisheries Technical Paper 225 rev. 1. Rome: Food and Agriculture Organization of the United Nations.

Rogers, G. L., R. Day, and A. Lim, eds. 1983. *Proceedings of the First International Conference on Warm Water Aquaculture—Crustacea.* Laie, HI: Brigham Young University Hawaii Campus.

Rosenberry, B., ed. 1991a. *Directory, shrimp farming in the Western Hemisphere.* San Diego, CA: Aquaculture Digest.

Rosenberry, B., ed. 1991b. *World shrimp farming.* San Diego, CA: Aquaculture Digest.

Spotts, D. 1983. Intensive shrimp culture in Japan. *Sea Frontiers* 29:272–7.

Taki, Y., J. H. Primavera, and J. A. Llobrera. 1984. *Proceedings of the First International Conference on the Culture of Penaeid Prawns/Shrimps.* Iloilo (Philippines): Aquaculture Dept., Southeast Asian Fisheries Development Center.

Villalon, J. R. 1991. *Practical manual for semi-intensive commercial production of marine shrimp.* TAMU-SG-91-501. Galveston: Texas Sea Grant Program.

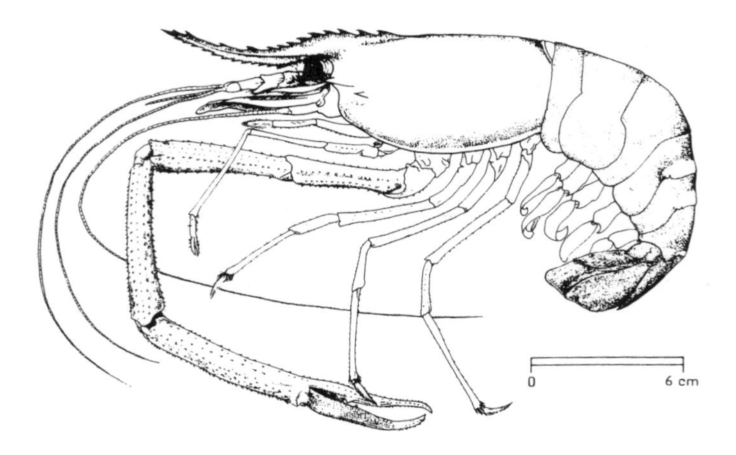

SCIENTIFIC AND COMMON NAMES: ***Macrobrachium rosenbergii,***
Also called freshwater prawn. **Giant river prawn**

DESCRIPTION AND DISTINCTIVE CHARACTERISTICS: The rostrum is well developed and has teeth. The carapace is spiny. The second pair of chelipeds are long, strong, and spiny with rather small claws.

HABITAT RANGE: Hawaii, Puerto Rico, and Florida to South Carolina, in warm waters. This native of Malaysia requires temperatures of 82–88 °F for optimum growth. It is widely transplanted worldwide wherever environmental conditions permit its survival.

REPRODUCTION AND DEVELOPMENT: The species spawns in brackish water, but the young move upstream into fresh water. A single large female may spawn 150,000–200,000 eggs per spawning, and may do so several times a year. The eggs are carried by the female for about 20 days; after 18–45 days as planktonic larvae they transform into the postlarval stage. This life cycle is said to be abbreviated when compared to the *Penaeus* shrimp since development is much more direct in *Macrobrachium,* involving fewer developmental stages.

AGE AND GROWTH: Growth is rapid in warm water; from egg to adult may require only about 2–3 months. Males may reach a weight of 23 oz.

FOOD AND FEEDING: Omnivorous, eating algae, aquatic worms, insects, small mollusks, crustaceans, plankton, and organic detritus.

PARASITES AND DISEASES: Bacteria (*Leucothrix* sp.), fungi (*Fusarium* sp.), protozoan (*Epistylus* sp.), digeneans. A disease called spontaneous necrosis also causes losses in prawn farms.

PREDATORS AND COMPETITORS: Birds and fishes; young prawns are eaten by dragonflies. Species is cannibalistic.

AQUACULTURAL POTENTIAL: Considerable understanding of the biology of this species, its rapid growth, and good local market demand in many areas make this a well-established farming operation. Prepared feeds are available. In the United States some buyer resistance has been experienced due to unfamiliarity with this exotic species, and there are marketing constraints. Hawaiian prawn output is now being sold on the U.S. mainland, and some success has been achieved in a summer-only crop in South Carolina. Giant river prawns reproduce readily in captivity, are much easier to grow than *Penaeus,* and will mate without rigid control of the environment. Larvae will begin feeding directly on brine shrimp. Although some salt water is required for rearing young juveniles, older juveniles and adults live in fresh water; this permits farming in areas away from expensive coastal property. *M. rosenbergii* has received the most attention as a candidate species in this genus; several other *Macrobrachium* species examined have been judged to be somewhat less desirable for commercial interest. Disadvantages of *Macrobrachium* include slower growth, longer larval life, and more aggressive behavior. The average size of individual river prawns decreases when reared at high densities, but greater production can be achieved. Differential growth rates even within a single population make harvesting and marketing more difficult. The edible portion of the river prawn (the tail) comprises 50 percent of its body weight, as compared to about 65 percent in penaeid shrimp.

REGIONS WHERE FARMED AND/OR RESEARCHED: Hawaii, South Carolina; worldwide in countries with suitable climate (see Ling and Costello 1979 for details).

REFERENCES

Goodwin, H. L., and J. A. Hanson. 1975. *The aquaculture of freshwater prawns (*Macrobrachium *species).* Waimanalo, Hawaii: Oceanic Institute.

Hanson, J. A., and H. L. Goodwin, eds. 1977. *Shrimp and prawn farming in the Western Hemisphere: State-of-the-art reviews and status assessments.* Stroudsburg, PA: Dowden, Hutchinson & Ross.

Ling, S. W., and T. J. Costello. 1979. The culture of freshwater prawns: A review. In *Advances in aquaculture; papers presented at the FAO Technical Conference on Aquaculture, 1976,* ed. T. V. R. Pillay and W. A. Dill, pp. 299–305. Farnham (U.K.): Fishing News Books.

New, M. B., and S. Singholka. 1985. *Freshwater prawn farming: A manual for the culture of Macrobrachium rosenbergii,* rev. ed. FAO Fisheries Technical Paper no. 225, rev. 1. Rome: Food and Agriculture Organization of the United Nations.

Sandifer, P. A., and T. I. J. Smith. 1985. Freshwater prawns. In *Crustacean and mollusk aquaculture in the United States,* ed. J. V. Huner and E. E. Brown, pp. 63–125. Westport, CT: AVI Publishing Co.

Shang, Y. C., and T. Fujimura. 1977. The production economics of freshwater prawn (*Macrobrachium rosenbergii*) farming in Hawaii. *Aquaculture* 11:99–110.

SCIENTIFIC AND COMMON NAMES: ***Penaeus vannamei,***
Also called Pacific blue shrimp, among other names. **Pacific white shrimp**

DESCRIPTION AND DISTINCTIVE CHARACTERISTICS: As noted in the preceding introductory section, the characteristics used to distinguish many of the *Penaeus* shrimp are very subtle. Scientists therefore describe the various species by using "identification keys," which are replete with technical terms. However, shrimp fishers and farmers can usually (with a bit of experience) distinguish several different species of *Penaeus* on sight, based on size, coloration, and overall appearance.

HABITAT RANGE: Northern Peru to Sonora, Mexico. Transplanted to many areas in the Greater Caribbean, South America, and the United States.

REPRODUCTION AND DEVELOPMENT: During mating the male attaches a packet of sperm cells to the female's sex organ soon after she molts. Between 500,000 and one million *demersal* (sinking) eggs are released free in the water upon fertilization. More than one spawning during the life of a shrimp is possible. In some areas, spawning takes place throughout the year. Drifting larval stages last about three weeks. Young juveniles live in nursery areas, which are usually large estuarine bays; as they mature, they move offshore to deeper saline waters where they spawn.

AGE AND GROWTH: Estimated to live about one year (perhaps as long as 20 months). Females may reach 9–11 in., males 7.5 in. total length.

FOOD AND FEEDING: Eat many different small plants and animals plus organic and inorganic detritus. Their diet includes dinoflagellates, nematodes, foraminiferans, algae, fish, snails, squids, clams, annelids, insects, and shrimp (cannibalistic).

PARASITES AND DISEASE: About 30 diseases, some infective, occur in cultured species, due to viruses, bacteria, fungi, and protozoans. Noninfective diseases are represented by gas-bubble disease, black gill syndrome, body cramp syndrome, muscle necrosis, and so on. Wild shrimp are also infected by a variety of larval stages of macroparasites that normally do not occur in artificially reared individuals.

PREDATORS AND COMPETITORS: A wide variety of finfish.

AQUACULTURAL POTENTIAL: Extremely high. This species was responsible for the rapid, unprecedented growth of shrimp farming in South America, especially Ecuador, where biological, economic, and social conditions in Ecuador were ideal for large-scale commercial shrimp culture. Some 80 percent of the shrimp farmed there are white leg shrimp.

REGIONS WHERE FARMED AND/OR RESEARCHED: Antigua, Bahamas, Dominican Republic, U.S. Virgin Islands, Guatemala, Brazil, Colombia, Ecuador, Peru, and others.

REFERENCES

Rogers, G. L., R. Day, A. Lim, eds. 1983. *Proceedings of the First International Conference on Warm Water Aquaculture–Crustacea.* Laie, HI: Brigham Young University Hawaii Campus.

Taki, Y., J. H. Primavera, and J. A. Llobrera, eds. 1984. *Proceedings of the First International Conference on the Culture of Penaeid Prawns/Shrimps.* Iloilo (Philippines): Aquaculture Dept., Southeast Asian Fisheries Development Center.

Echinoderms

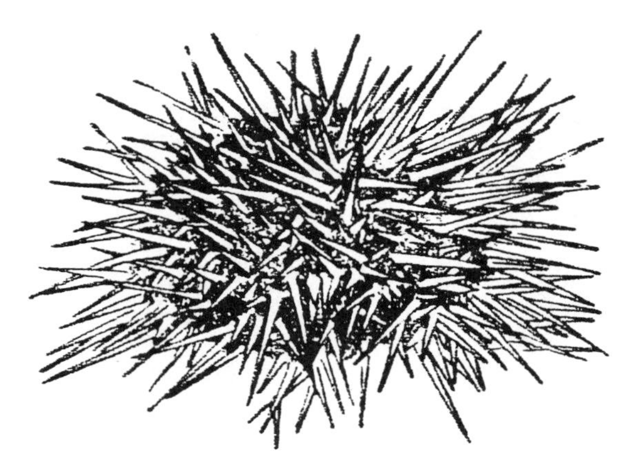

Sea Urchin

SCIENTIFIC AND COMMON NAMES: ***Strongylocentrotus franciscanus*, Red sea urchin (U.S.)**
In Japan, *Hemicentrotus pulcherrimus, Pseudocentrotus depressus,* and *Anthocidaris crassispina* are the three major species; five minor species are also used for food.

DESCRIPTION AND DISTINCTIVE CHARACTERISTICS: Shaped like a tangerine, the *test* (hard shell) is covered with spines of varying lengths and thickness.

HABITAT RANGE: Red sea urchin extends from Baja, California, northward to Kodiak, Alaska, and thence across the Pacific to Japan. (The green sea urchin, *S. droebachiensis,* is found in the northern part of both U.S. coasts.) Urchins prefer rocky ground and avoid rough waves.

REPRODUCTION AND DEVELOPMENT: Sexual maturity is reached at between 1 and 2 years of age. The planktonic larval stage is called *echinopluteus.*

AGE AND GROWTH: Fertilization to settlement lasts about six to eight weeks.

FOOD AND FEEDING: Eat algae, including the large *Macrocystis* (kelp), and plankton; apparently feed most actively at temperatures of 43–77 °F.

PARASITES AND DISEASE: Lesions caused by bacteria occur naturally on the body wall of some species of sea urchins. This communicable disease causes color changes in the epidermis, and spines and other appendages are lost; with heavy infection, mortality may ensue. A protozoan disease causes loss of muscle function in the tube feet. Algae (diatoms) can cause death by disintegration of the epidermis. Turbellarian worms parasitize the digestive tract and coelomic cavity. Occasionally, mass mortalities of unknown causes occur in some urchin populations.

PREDATORS AND COMPETITORS: Rock crabs, starfish, sea stars, spiny lobsters, sharks, wolfeels, California sheephead, sea otters, and sea gulls are predators. Abalone compete with sea urchins for space.

AQUACULTURAL POTENTIAL: Both male and female sea urchins are harvested for their bright yellow gonads, which are marketed as "roe" (*uni* in Japanese). (The word "roe" usually refers to eggs of aquatic fishes when still enclosed in the ovarian membrane.) Increased demand for sea urchin roe suggests that species of *Strongylocentrotus* will be more popular in the future. They feed low on the food web and grow rapidly.

REGIONS WHERE FARMED AND/OR RESEARCHED: Hawaii; Japan.

REFERENCES

Kato, S., and S. C. Schroeter. 1985. Biology of the red sea urchin, *Strongylocentrotus franciscanus,* and its fishery in California. *Marine Fisheries Review* 47(3):1–20.

Mottet, M. G. 1976. *The fishery biology of sea urchins in the family Strongylocentrotidae.* State of Washington, Department of Fisheries Technical Report no. 20. Olympia: Washington, Department of Fisheries.

Takagi, K. 1986. Aspects of the sea urchin fishery in Japan. *Proceedings of the Gulf and Caribbean Fisheries Institute* 37:41–53.

Mollusks

Mollusks have soft bodies enclosed in a fleshy mantle. Most members of the phylum Mollusca have an external hard, calcareous shell. The unsegmented body has a head, a dorsally located visceral mass, and a large ventral foot that enables very limited movement. Internal shells permit squids and octopuses (cephalopods) a streamlined shape that enables them to swim rapidly.

Mollusks are found in the sea from the shoreline to depths of 35,000 ft. Most eat seaweed and microorganisms (i.e., are filter feeders or plankton feeders), except for squids and octopuses, which eat other mollusks, crustaceans, and fishes. Some gastropods (oyster drills, tulip snails) are also carnivorous. They are eaten by a wide variety of predators, including crustaceans, notably crabs, lobsters, other Mollusca (drills, octopuses), and finfishes.

Most mollusks serve as intermediate hosts for parasitic worms; certain well-studied ones, such as oysters, are also parasitized by a wide range of other organisms, from viruses to helminths. Spaces within the shells provide convenient shelter for many small crustaceans. Mollusks that burrow (e.g., clams) have fewer symbionts present on the exterior of their shells than those that do not (abalones, oysters, etc.). These symbionts provide a suitable substrate for many commensal encrusting organisms, algae, tube worms, boring sponges, and other mollusks.

Bivalve mollusks can cause illness in humans in three different ways:

1. Some people are allergic to shellfish.
2. Bivalves from polluted areas contain pathogenic bacteria and viruses.
3. Filter-feeding mollusks that have been feeding on certain dinoflagellates (protozoans that live in the plankton) or toxic algae can cause paralytic shellfish poisoning (PSP) when eaten.

It is very difficult to determine the presence of these toxic planktonic organisms in the water where mollusks live: Usually the first sign is an outbreak of the potentially fatal PSP among persons eating the shellfish. Naturally, coverage of such occurrences by the news media severely hurts the capture and culture fisheries markets.

Shellfish from areas polluted by human sewage can be cleansed or *depurated* (purified) by placing them in clean seawater for an extended period. This procedure, using special depuration plants, is rather common in some European countries. In the United States, oysters can be taken from polluted areas and *relayed* (placed in a natural area) with clean water, making them suitable for human consumption (after the proper period).

REFERENCES

Hahn, K. O., ed. 1988. *Handbook of culture of abalone and other marine gastropods*. Boca Raton, FL: CRC Press.

Huner, J. V., and E. E. Brown, eds. 1985. *Crustacean and mollusk aquaculture in the United States*. Westport, CT: AVI Publishing Co.

Menzel, W., ed. 1991. *Estuarine & marine bivalve mollusk culture*. Boca Raton, FL: CRC Press.

Otwell, S., and G. E. Rodrick, eds. 1991. *Molluscan shellfish depuration*. Boca Raton, FL: CRC Press.

Shumway, S. E. 1989. Toxic algae: A serious threat to shellfish aquaculture. *World Aquaculture* 20(4): 65–74.

Bivalves

Clams

Clams are the most commercially important species of bivalve in the United States: Commercial landings of all species of clams are over three times greater than those of oysters. They are also sought by sportfishers. Clams are found in a wide geographic range, on virtually every coastline in the Northern Hemisphere, including all along the Atlantic and Pacific coasts.

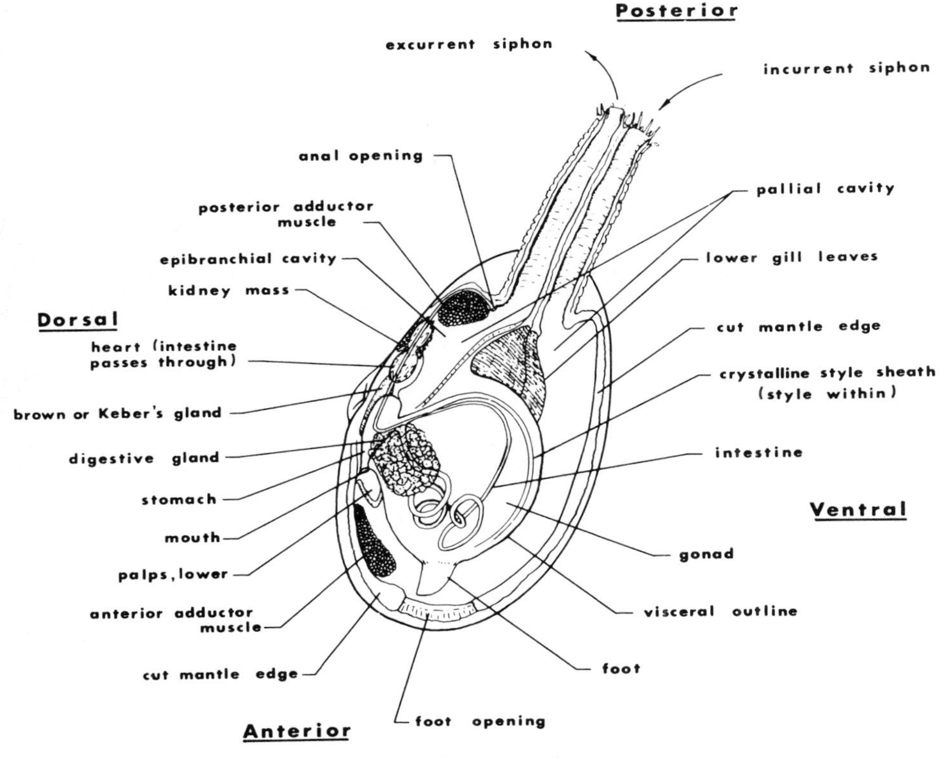

Clam descriptive terminology.

As in all bivalves, the two shells (or *valves*) are opened and closed by powerful muscles; likewise, the sexes are separate (distinct). Clams feed by filtering algae and diatoms over the gills and passing the food to the mouth via hairlike *cilia*. Limited movement is achieved by extending a large foot to propel the animal across the seafloor.

Species commercially and recreationally valuable in the United States are surf clams (*Spisula solidissima*), hard clams (*Mercenaria mercenaria*), and soft-shell clams (*Mya arenaria*). A good overview of the status of North American clam mariculture can be found in the references listed below.

Clams are not as desirable for farming as are oysters: They have a slower growth rate, a lower per-pound market price, and seed stock that is difficult to collect in sufficient quantity. Clam farmers generally experience high clam mortality due to predation by crabs.

In Florida, the feasibility of culturing angelwing clam, *Cyrtopleura costata*, is being tested. This deep-burrowing clam grows rapidly, has a fine flavor, and apparently would support a large market. Early life stages have been reared successfully at Harbor Branch Oceanographic Institution, Fort Pierce. Techniques to produce high survival in field grow-out remain to be developed.

REFERENCES

Manzi, J. J. 1991. Clams, cockles and arkshells. In *Production of aquatic animals; crustaceans, amphibians and reptiles*, ed. C. E. Nash, pp. 139–59.

Manzi, J. J., and M. Castagna, eds. 1989. *Clam mariculture in North America*. Developments in Aquaculture and Fisheries Science, vol. 19. Amsterdam: Elsevier.

SCIENTIFIC AND COMMON NAMES: ***Tridacna gigas*, Giant clam**
Related: *T. derasa, T. squamosa, Hippopus hippopus*

DESCRIPTION AND DISTINCTIVE CHARACTERISTICS: An extremely large clam, believed to be the largest known bivalve mollusk; reaches over 4 ft and 660 lb. Possesses a brightly colored *mantle* (siphonal tissue), which harbors numerous symbiotic dinoflagellates (zooxanthellae).

HABITAT RANGE: Reefs in clear, shallow waters of the tropical Indo-Pacific.

REPRODUCTION AND DEVELOPMENT: Highly fecund: Some 500 million eggs may be released in a single day. Like all tridacnids, these clams are *hermaphroditic* (have both sexes in the same individual), become sexually mature at 3–3.5 years of age, and spawn during the summer. Larval drifting lasts two or three weeks.

AGE AND GROWTH: Believed to live about 30 years. Time required to reach 4-in. length is unknown, but some 12 years are required thereafter to reach about 32 in. May reach 20 in. and 13 lb. in 5–7.5 years.

FOOD AND FEEDING: Dinoflagellates are believed to contribute substantially to the nutrition of these clams and may also be involved in calcification. Symbiotic brown algae (zooxanthellae) produce carbohydrates directly into clam tissues. Nitrogenous wastes of the clam are consumed by the algae.

PARASITES AND DISEASE: Numerous bacteria and bacterial diseases (mainly *Vibrio* and *Pseudomonas*), a protozoan (*Perkinsus*), a nematode, and a trematode species have been reported for different species of giant clams from various locations.

PREDATORS AND COMPETITORS: Giant clams are subject to heavy predation in the field when small. Tridacnids under about 3 years of age are eaten by octopods, hermit crabs, carnivorous snails, rays, fishes, and turtles.

AQUACULTURAL POTENTIAL: Strong markets exist in Japan, Taiwan, and the People's Republic of China. Considerable area is available for culture, with roughly 100 tons per acre suggested as the possible annual yield. The fact that it takes approximately three years to reach market size is a possible disadvantage to profitable culture, but the lack of feed costs is an advantage. These clams can be spawned and raised in hatcheries. Their attractive shells are a bonus.

REGIONS WHERE FARMED AND/OR RESEARCHED: Papua New Guinea; Palau, Caroline Islands; Orpheus Island Research Station, Australia; Hawaii. Introduced to Bonaire and St. Lucia.

REFERENCES

Braley, R. D. 1988. Farming the giant clam. *World Aquaculture* 20(1): 6–17.

Copland, J.W., and J. S. Lucas, eds. 1988. *Giant clams in Asia and the Pacific.* Canberra, Australia: Australian Centre for International Agricultural Research.

Munro, J. L., and G. A. Heslinga. 1983. Prospects for the commercial cultivation of giant clams (Bivalvia: Tridacnidae). *Proceedings of the Gulf and Caribbean Fisheries Institute* 35:122–34.

SCIENTIFIC AND COMMON NAMES: ***Asaphis deflorata,* Tiger lucine**
***Codakia orbicularis,* Great white lucine**

DESCRIPTION AND DISTINCTIVE CHARACTERISTICS: Approximately orbicular in outline. Shell valves are white, thick, and strong, with heavy radial ribs on the outside.

HABITAT RANGE: In nutrient-poor waters from Bermuda and Florida, throughout the Caribbean and southern Gulf of Mexico, to as far south as Brazil.

REPRODUCTION AND DEVELOPMENT: Spawning continuous from late spring through late fall, with spring and fall peaks of activity.

AGE AND GROWTH: May reach a shell length of 3.5 in.

FOOD AND FEEDING: May have chemotropic capabilities through a symbiotic relationship with sulphur-fixing bacteria within their gill tissues.

PARASITES AND DISEASE: No record.

PREDATORS AND COMPETITORS: Are eaten by crabs.

AQUACULTURAL POTENTIAL: Apparently rather limited; they are little used as human food by capture fisheries today. Tiger lucine shells have been found in shell middens of Meso-Indians living on Antigua about 1777 (±90) B.C.

REGIONS WHERE FARMED AND/OR RESEARCHED: Due to their chemotropic abilities, lucines may be viable candidates in the plankton-poor Caribbean.

REFERENCES

Berg, C. J. 1984. Aquaculture potential of Bermudian bivalve mollusks. In *Assessment of the potential for aquaculture in Bermuda,* ed. T. D. Sleeter, pp. 77–89. Bermuda Biological Station Special Publication no. 27. Ferry Reach: Bermuda Biological Station.

Berg, C. J., and P. Alatalo. 1982. Mariculture potential of shallow-water Bahamian bivalves. *Journal of the World Mariculture Society* 13:294–300.

Berg, C. J., and P. Alatalo. 1984. Potential of chemosynthesis in molluscan mariculture. *Aquaculture* 39:165–79.

SCIENTIFIC AND COMMON NAMES: ***Mercenaria mercenaria,* Hard clam** Also called round clam or quahog.

DESCRIPTION AND DISTINCTIVE CHARACTERISTICS: Thick, hard, equal-sized shells (*valves*) with ridges. Two equal-sized adductor muscles; short siphons.

HABITAT RANGE: Distributed from the Gulf of St. Lawrence through the Gulf of Mexico to the Yucatán Peninsula; from the intertidal zone to depths of 50 ft.

REPRODUCTION AND DEVELOPMENT: Most spawn when 1 year old. Eggs are pumped out of the siphon, and fertilization is external. Young stages settle on the bottom after about 12 days.

AGE AND GROWTH: May reach 3.5 in. after about five years, and 5.5 in. if live to 20–25 years.

FOOD AND FEEDING: Eat plankton, mostly phytoplankton, drawing food into the mouth through short siphons.

PARASITES AND DISEASES: Bacteria, fungi, trematodes, nemerteans, nematodes, mollusks, and crustaceans. Hard clams may develop noninfectious diseases (e.g., cancer, gas bubble disease, and tumors).

PREDATORS AND COMPETITORS: To control predation on very small clams in hatcheries and nurseries, in-take water systems require fine-mesh filters. Young hard clams are subject to very high predation by crabs, moon snails, starfish, whelks, gulls, and oyster catchers; even large, heavy-shelled clams are eaten by some of these predators. Recently the big-clawed snapping shrimp, *Alpheus heterochaelis,* was found to be a hard-clam predator. Live toadfish placed in clam cages will eat crabs that enter to prey on the clams, but will not harm the clams themselves.

AQUACULTURAL POTENTIAL: Hard-clam aquaculture has been researched for over 80 years, yet satisfactory technology is still not clearly defined. High cost is the culprit in raising hard clams to market size for profit: Much attention must be devoted to their care, protection from predators, and cleanliness.

REGIONS WHERE FARMED AND/OR RESEARCHED: Indian River, Florida; Virginia Institute of Marine Science, Gloucester Point; South Carolina.

REFERENCES

Castagna, M., and J. N. Kraeuter. 1981. *Manual for growing the hard clam* Mercenaria. Special Report in Applied Marine Science and Ocean Engineering, no. 249. Gloucester Point, VA: Virginia Institute of Marine Science.

Hartman, M., C. E. Epifanio, G. Pruder, and R. Srna. 1974. Farming the artificial sea: Growth of clams in a recirculating seawater system. *Proceedings of the Gulf and Caribbean Fisheries Institute* 26:59–74.

McHugh, J. L., M. W. Sumner, P. J. Flagg, D. W. Lipton, and W. J. Behrens. 1982. *Annotated bibliography of the hard clam* (Mercenaria mercenaria). NOAA Technical Report NMFS SSRF-756. Seattle: National Marine Fisheries Service.

McHugh, J. L., and M. W. Sumner. 1988. *Annotated bibliography II of the hard clam* Mercenaria mercenaria. NOAA Technical Report NMFS 68. Seattle: National Marine Fisheries Service.

Manzi, J. J., and M. Castagna, eds. 1989. *Clam mariculture in North America.* Developments in Aquaculture and Fisheries Science, vol. 19. Amsterdam: Elsevier.

SCIENTIFIC AND COMMON NAMES: ***Tapes japonica*, Manila clam**
Also called Japanese littleneck clam.

DESCRIPTION AND DISTINCTIVE CHARACTERISTICS: A small clam with oval valves, both of which have well-defined radiating ribs and concentric rings. Color varies from yellow to buff, and there may be wavy brown or black bands or blotches on the shells.

HABITAT RANGE: The Manila clam was introduced to Pacific Northwest waters (Washington State and British Columbia) with shipments of Pacific oyster seed from Japan during the 1930s and 1940s. It is abundant on public beaches in Washington State, in bays and inlets of southern Puget Sound; also found in Humboldt and San Francisco bays, California.

REPRODUCTION AND DEVELOPMENT: In Washington State, spawning occurs in June and July; in Japan, it takes place in spring and autumn.

AGE AND GROWTH: Larvae are planktonic for about three weeks. Market size of 15–25 clams per pound may be reached at age 2–3 years. Maximum size is about 3 in. at widest point.

FOOD AND FEEDING: Filter feeder.

PARASITES AND DISEASE: A larval trematode has been found in this clam from Korean waters.

PREDATORS AND COMPETITORS: Eaten by moon snails, crabs (including the Dungeness crab), a variety of fish (including rock sole, English sole, starry flounder, and pile perch), starfish, and scooter ducks. Probably commensal with pinnotherid crabs.

AQUACULTURAL POTENTIAL: Fine flavor. Difficult to grow out in unenclosed areas because of high levels of predation.

REGIONS WHERE FARMED AND/OR RESEARCHED: Puget Sound, Washington; British Columbia; Japan.

REFERENCES

Anderson, G. J., M. B. Miller, and K. K. Chew. 1982. *A guide to Manila clam aquaculture in Puget Sound.* Washington Sea Grant Technical Report WSG 82-4. Seattle: Washington Sea Grant Program.

Chew, K. K. 1989. Manila clam biology and fishery development in western North America. In *Clam mariculture in North America,* ed. J. J. Manzi and M. Castagna, pp. 243–61. Developments in Aquaculture and Fisheries Science, vol. 19. Amsterdam: Elsevier.

Robinson, A. M., and W. P. Breese. 1984. Gonadal development and hatchery rearing techniques for the Manila clam *Tapes philippinarum* (Adams and Reeve). *Journal of Shellfish Research* 4 :161–3.

SCIENTIFIC AND COMMON NAMES: ***Mya arenaria,* Soft-shell clam** Also called longneck and steamer clam.

DESCRIPTION AND DISTINCTIVE CHARACTERISTICS: The two valves are irregular ellipsoids about equal in size. The outer shell surface is covered with rough striations, some more pronounced than others. "Annual" rings or marks appear on the shell. Hinge teeth are prominent.

HABITAT RANGE: Sub- and intertidal waters from Labrador to North Carolina on the East Coast and in several scattered locations on the West Coast of the United States, including San Francisco; Northern Norway to Bay of Biscay, France.

REPRODUCTION AND DEVELOPMENT. Sexual maturity may be reached at about 1 year of age; each female releases millions of eggs. Larvae drift, carried by currents, then metamorphose and settle to the bottom, moving short distances; as they get older, they dig a permanent burrow and remain there.

AGE AND GROWTH: Soft-shell clams in Chesapeake Bay attain a shell length of about 2 in. at age 1.5–2 years; may be age 5–6 before reaching this same size in Maine.

FOOD AND FEEDING: Eat microscopic plankton by filtration, predominantly dinoflagellates.

PARASITES AND DISEASE: Viruses, bacteria, protozoans, trematodes, nemerteans, nematodes, mollusks, crustaceans; neoplastic "tumors."

PREDATORS AND COMPETITORS: Heavily preyed upon by numerous crab species (esp. green crab); also drills, ducks, swans, flounders and other bottom-feeding fish, and sea gulls.

AQUACULTURAL POTENTIAL: A very strong market exists for this clam, and there is a wealth of hatchery technology on clam rearing; however, the predator problem is still a major deterrent to the growing out of unprotected clams in the wild.

REGIONS WHERE FARMED AND/OR RESEARCHED: Virginia Institute of Marine Science, Gloucester Point; South Carolina.

REFERENCES

Huner, J. V., and E. E. Brown, eds. 1985. *Crustacean and mollusk aquaculture in the United States*. Westport, CT: AVI Publishing Co.

Manzi, J. J., and M. Castagna, eds. 1989. *Clam mariculture in North America*. Developments in Aquaculture and Fisheries Science, vol. 19. Amsterdam: Elsevier.

COMMON AND SCIENTIFIC NAME: ***Spisula solidissima,* Surf clam**
Also called sea, bar, beach, and skimmer clam.

DESCRIPTION AND DISTINCTIVE CHARACTERISTICS: Smooth oval valves with small growth lines. Valves do not close tightly. Coloration is cream to tan.

HABITAT RANGE: Along the eastern coast of North America, from Nova Scotia to South Carolina. Open ocean below low water line, down to 100 ft. on sandy bottom. Surf clams burrow actively in sand. Maximum temperature is estimated at 86 °F, based on a related species. Minimum salinities for survival are 16 ppt (parts per thousand) for larvae and about 12 ppt for adults.

REPRODUCTION AND DEVELOPMENT: In New Jersey, spawning takes place in early summer, with a lesser spawning in early fall.

AGE AND GROWTH: Rapid growth of 2 in. within a year (May–October). May reach length of 7 in.

FOOD AND FEEDING: Filter feeder.

PARASITES AND DISEASE: Protozoans, trematodes, cestodes, and nematodes.

PREDATORS AND COMPETITORS: Predators include two species of moon snail, a boring snail, and fishes (haddock and cod). Little is known of inter- and intra-specific competition involving surf clams.

AQUACULTURAL POTENTIAL: Rapid growth, market potential, and an ability to withstand changing environmental conditions suggest that the surf clam should be considered as a candidate for commercial aquaculture. In addition, seed stock rearing technology is well established, and production is dependable. Grow-out to market size (2 in.) has been carried out in cages and on land in raceways, but large-scale commercial production has yet to be effected.

REGIONS WHERE FARMED AND/OR RESEARCHED: Throughout its geographic range, with most research in Milford, Connecticut.

REFERENCES

Fay, C. W., R. J. Neves, and G. B. Pardue. 1983. *Species profiles: Life histories and environmental requirements of coastal fishes and invertebrates (Mid-Atlantic)-surf clam.* U.S. Fish and Wildlife Service FWS/OBS-82/11.13. U.S. Army Corps of Engineers TR EL-82-4. Slidell, LA: National Coastal Ecosystems Team, U.S. Fish and Wildlife Service.

Manzi, J. J., and M. Castagna, eds. 1989. *Clam mariculture in North America.* Developments in Aquaculture and Fisheries Science, vol. 19. Amsterdam: Elsevier.

Winter, J. E., J. E. Toro, J. M. Navarro, G. S. Valenzuela, and O. R. Chaparro. 1984. Recent developments, status and prospects of molluscan aquaculture on the Pacific Coast of South America. *Aquaculture* 39(1–4):95–134.

Yancey, R. M., and W. R. Welch. 1968. *The Atlantic coast surf clam, with a partial bibliography.* U.S. Fish and Wildlife Service Circular no. 288. Washington, DC: U.S. Fish and Wildlife Service.

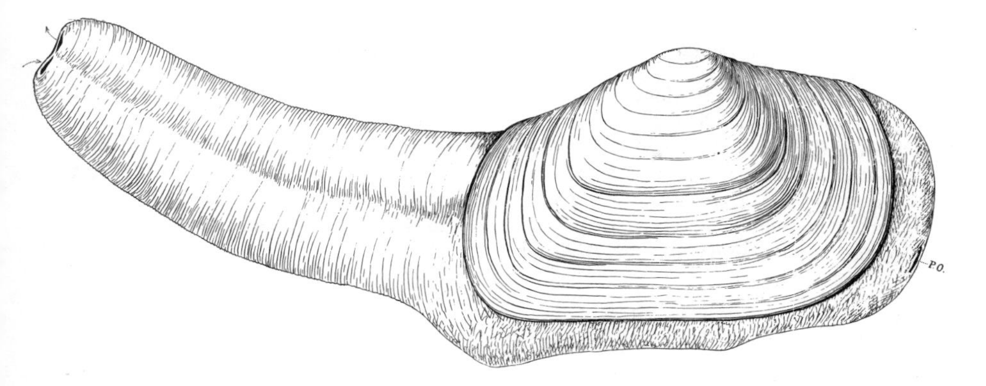

SCIENTIFIC AND COMMON NAMES: ***Panope abrupta (generosa),* Geoduck**

DESCRIPTION AND DISTINCTIVE CHARACTERISTICS: The largest of the burrowing clams, reaching a length of 7 in. and a weight of 10 lb. The white shells show growth lines.

HABITAT RANGE: In the intertidal zone from below mean lower low water to depths of 200 ft. May burrow as deep as 3 ft., usually in soft sand–mud mixtures. Are abundant in Puget Sound, with a total range from Alaska to the Baja Peninsula.

REPRODUCTION AND DEVELOPMENT: Sexual maturity reached at about age 3.

AGE AND GROWTH: There is considerable regional growth variation: In some areas, total weight may reach 2 lb. in as little as three years; in others, geoducks may not reach this weight until age 8–10.

FOOD AND FEEDING: The geoduck is a filter feeder and lives on small plankton, benthic diatoms, and detritus.

PARASITES AND DISEASE: A flagellate (protozoan) has been found in hatchery-reared larval geoducks. Heavy infections killed the larvae by penetrating the mantle and proliferating in the coelom.

PREDATORS AND COMPETITORS: Larvae serve as food for a variety of fish and invertebrates. Juveniles and adults, because of their large size and burying behavior, are relatively free from predators.

AQUACULTURAL POTENTIAL: Geoduck larvae have been cultured successfully, and rearing techniques have been developed to raise them to a size for planting in the wild. Washington State leases bay bottom in Puget Sound to private citizens for clam farming: Ten such leases, covering approximately 1,200 acres of bay bottom, have been granted for farming geoducks.

REGIONS WHERE FARMED AND/OR RESEARCHED: Seattle, Washington.

REFERENCES

Cheney, D. P., and T. F. Mumford, Jr. 1986. *Shellfish & seaweed harvests of Puget Sound.* Seattle: Washington Sea Grant Program.

Goodwin, L., and B. Pease. 1987. *The distribution of geoduck* (Panope abrupta) *size, density, and quality in relation to habitat characteristics such as geographic area, water depth, sediment type, and associated flora and fauna in Puget Sound, Washington.* State of Washington, Department of Fisheries Technical Report no. 102. Olympia, WA: Washington Department of Fisheries, Shellfish Division.

Mussels

Marine bivalve mussels used as human food have thin shells colored dark purple to black. They are distributed in temperate waters, and differ from clams in that their *byssus* threads, which serve as a holdfast, remain throughout their life. They are especially popular in Spain, Holland, France, Italy, and parts of South America, as well as in the Philippines.

Evidence is still being gathered to evaluate the advisability of farming the California mussel (*Mytilus californianus*) into Puget Sound, Washington: It grows and survives well, but may have drawbacks. In one experiment in Oregon, *M. californianus* grew more rapidly than the blue mussel (*M. edulis*), and had twice as much dry meat.

The perna mussel (*Perna perna*), which has been farmed in Venezuela, is cosmopolitan in tropical and subtropical coastal waters.

REFERENCES

Chalermwat, K., and R. A. Lutz. 1989. Farming the green mussel in Thailand. *World Aquaculture* 20(4):41–6.

Lutz, R. A., ed. 1980. *Mussel culture and harvest: A North American perspective.* Developments in Aquaculture and Fisheries Science, vol. 7. Amsterdam: Elsevier.

Skidmore, D., and K. K. Chew. 1985. *Mussel aquaculture in Puget Sound.* Seattle: Washington Sea Grant Program, University of Washington.

Yamada, S. B., and J. B. Dunham. 1989. *Mytilus californianus*, a new aquaculture species? *Aquaculture* 81:275–84.

SCIENTIFIC AND COMMON NAMES: ***Mytilus edulis,* Blue mussel**
Also called sea mussel.

DESCRIPTION AND DISTINCTIVE CHARACTERISTICS: Bivalve mollusk characterized by a byssus secreted from a gland at base of foot. The shell is thin and black to bluish-black.

HABITAT RANGE: Widely distributed in north temperate waters.

REPRODUCTION AND DEVELOPMENT: Each female spawns 5–12 million eggs annually. A ciliated larva is formed about four hours after fertilization; after 10 weeks, all organs are present. Adults are sessile.

AGE AND GROWTH: About 1 in. per year for the first three years.

FOOD AND FEEDING: Eat plankton, small diatoms (29 spp.), protozoans (9 spp.), and detritus. Currents, set up by gills, draw floating food into mouth.

PARASITES AND DISEASE: Parasitized by protozoans (several species), a mesozoan, trematodes, and crustaceans (copepods and pinnotherid crabs). Diseases include hemic neoplasmic disorders.

PREDATORS AND COMPETITORS: Mussels serve as food for crabs (especially blue crabs), starfish, whelks, birds, and mammals (rats on intertidal zone, muskrats, and walruses).

AQUACULTURAL POTENTIAL: Mussel consumption is very low in North America, but could be increased markedly by a strong, though expensive, consumer-education program. Demand is presently high in South America and Europe, especially Spain. This tasty bivalve can be produced in great abundance in relatively small areas using some of the off-the-bottom rearing techniques. Feeding is not required for grow-out because mussels consume the plankton that occur naturally in seawater. As pointed out earlier (in the introductory section "Mollusks"), filter-feeding mollusks may accumulate toxins from some phytoplankton species and, when eaten by humans, cause serious illness (PSP). In 1987, an outbreak of amnesic shellfish poisoning in Prince Edward Island, Canada, poisoned 129 people and caused two deaths.

REGIONS WHERE FARMED AND/OR RESEARCHED: Maine and Washington; Spain, Holland, and France; also Australia (subspecies *M. edulis planulatus*).

REFERENCES

Hurlburt, C. G., and S. W. Hurlburt. 1974. *Blue gold—mariculture of the edible blue mussel,* Mytilus edulis. Private publication.

Lutz, R. A. 1985. Mussel aquaculture in the United States. In *Crustacean and mollusk aquaculture in the United States,* ed. J. V. Huner and E. E. Brown, pp. 311–63. Westport, CT: AVI Publishing Co.

SCIENTIFIC AND COMMON NAMES: ***Perna viridis*, Green mussel**
Formerly *Mytilus smaragdinus.*

DESCRIPTION AND DISTINCTIVE CHARACTERISTICS: Presence of primary lateral hinge teeth, absence of anterior adductor muscle, and the recurrent loop of the midgut on the left side of the stomach.

HABITAT RANGE: Full-strength seawater; shallow coastal waters in India, Malaysia, Philippines, Singapore, Thailand, Pakistan, China, New Zealand, and Africa.

REPRODUCTION AND DEVELOPMENT: Sexes are separate (distinct). Females attain sexual maturity at 1 in., then spawn year-round. A trochopore-stage larva develops from the egg after 6–8 hr; after 16–18 hr the first larval shell is secreted while still planktonic. Metamorphosis is reached in 10–30 days.

AGE AND GROWTH: May reach market size in six or eight months in Malaysia; in India lengths of 4, 5.25, and 6.25 in. were reached at the end of the first, second, and third years.

FOOD AND FEEDING: Filter feeder.

PARASITES AND DISEASE: A parasitic copepod, *Anthessius mytilicola.*

PREDATORS AND COMPETITORS: The most important predator of the green mussel is a crab, *Scylla serrata.* Starfish prey heavily on mussels that are not cultured off the bottom. Some fouling organisms attach to the shells of mussels (e.g., barnacles, filamentous algae, bryozoans, and tunicates); these may reduce the rate of growth and/or affect the taste of the mussel.

AQUACULTURAL POTENTIAL: Mussels are a well-known seafood with a fine flavor; however, they may feed on toxic dinoflagellates poisonous to humans. Setting up a farm to raise *P. viridis* is a low-cost proposition, usually requiring bamboo stakes and somewhat more for raft culture, but still using relatively inexpensive materials that are close at hand. By employing the kind of spat collector appropriate to a given location, and at a time when spat is plentiful, abundant seed can be produced for a farm at low cost.

REGIONS WHERE FARMED AND/OR RESEARCHED: Throughout its geographic range, especially in the Philippines and Thailand.

REFERENCES

Chalermwat, K., and R. A. Lutz. 1989. Farming the green mussel in Thailand. *World Aquaculture* 20(4):41–6.

Vakily, J. M. 1989. *The biology and culture of mussels of the genus* Perna. ICLARM Studies and Reviews, no. 17. Manila (Philippines): International Center for Living Aquatic Resources Management.

..ters

The oyster is a sedentary mollusk with two hard shells, or *valves,* that are attached by a hinge and held together snugly by a strong muscle. This armor of calcium carbonate shields the fleshy part of the oyster against predators and adverse environmental conditions.

In general, oyster species fall into two generic groups (Table 3.1): flat (*Ostrea*) and cup-shaped (*Crassostrea*). In all oysters, the two valves are dissimilar; in *Ostrea,* which is elongate from hinge end to shell margin, the right (upper) valve is flat and the left (lower) valve deeply cupped. *Ostrea* reproduces by releasing larvae free in the water, and *Crassostrea* by releasing eggs. *Crassostrea* can tolerate higher salinities and turbidity than *Ostrea.*

Oysters feed by pumping water between their valves, filtering out microorganisms in the water by specially adapted hairlike structures (*cilia*) on the gills, which also serve to move food through the mouth. They have many predators, including gastropods (oyster drills, whelks), starfish, flatworms, crabs, and fish. Oysters, in common with many other mollusks, are most vulnerable to predation when they are small and their shells are very thin.

Both Maine and Nova Scotia have limited production of the imported European oyster, *Ostrea edulis,* which is better suited for the cold water temperatures in those areas than is the American oyster.

Table 3.1. Important Food Oysters

Genus and Species	Common Name	Geographic Distribution	
		Native	Transplant
Flat oysters (*Ostrea*)			
O. edulis	Flat oyster or native European oyster	Atlantic Coast of Europe	Maine, Maritime Provinces of Canada
O. lurida	Olympia oyster	Alaska to Lower California	
Cup-shaped oysters (*Crassostrea*)			
C. brasiliana	—	Brazil	—
C. commercialis	Sydney rock oyster	Australia	Hawaii
C. gigas	Japanese or Pacific oyster	Japan	Alaska to California
C. paraibanensis	—	Northeastern Brazil	—
C. rhizophorae	Mangrove oyster	Southern Florida to Brazil	—
C. virginica	American or Eastern oyster	Atlantic Coast of N. America, Gulf of Mexico	California to Washington, Hawaii

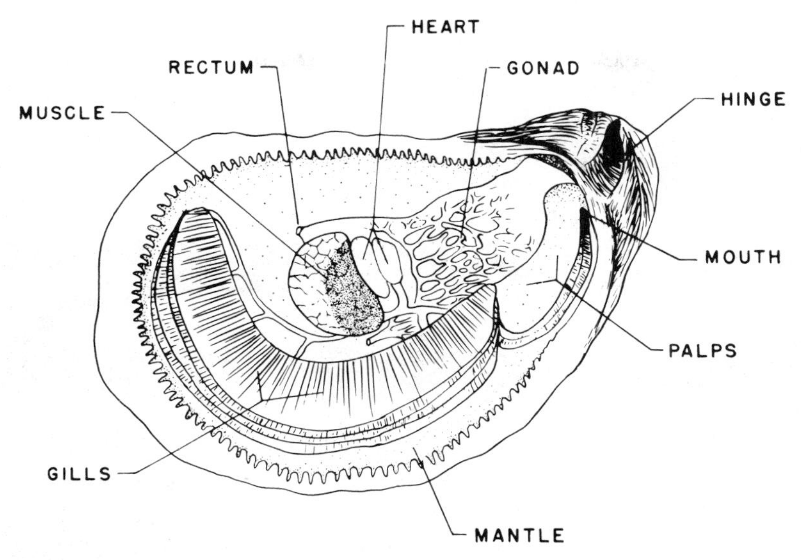

Oyster descriptive terminology.

REFERENCES

Angell, C. L. 1986. *The biology and culture of tropical oysters.* ICLARM Studies and Reviews, no. 13. Manila (Philippines): International Center for Living Aquatic Resources Management.

Breisch, L. L., and V. S. Kennedy. 1980. *A selected bibliography of worldwide oyster literature.* University of Maryland Sea Grant publication UM-SG-TS-80-11. College Park: University of Maryland Sea Grant Program.

Burrell, V. B. 1985. Oyster culture. In *Crustacean and mollusk aquaculture in the United States,* ed. J. V. Huner and E. E. Brown, pp. 235–73. Westport, CT: AVI Publishing Co.

Galtsoff, P. S. 1964. The American oyster *Crassostrea virginica* Gmelin. *Fishery Bulletin* 64:1–480.

Hidu, H., and R. E. Lavoie. 1991. The European oyster, *Ostrea edulis* L., in Maine, U.S., and eastern Canada. In *Estuarine and marine bivalve mollusk culture,* ed. W. Menzel, pp. 35–46. Boca Raton, FL: CRC Press.

Quayle, D. B. 1988. *Pacific oyster culture in British Columbia.* Canadian Bulletin of Fisheries and Aquatic Sciences no. 218. Ottawa: Dept. of Fisheries and Oceans.

Shaw, W. N. 1969. The past and present status of off-bottom oyster culture in North America. *Transactions of the American Fisheries Society* 98:755–61.

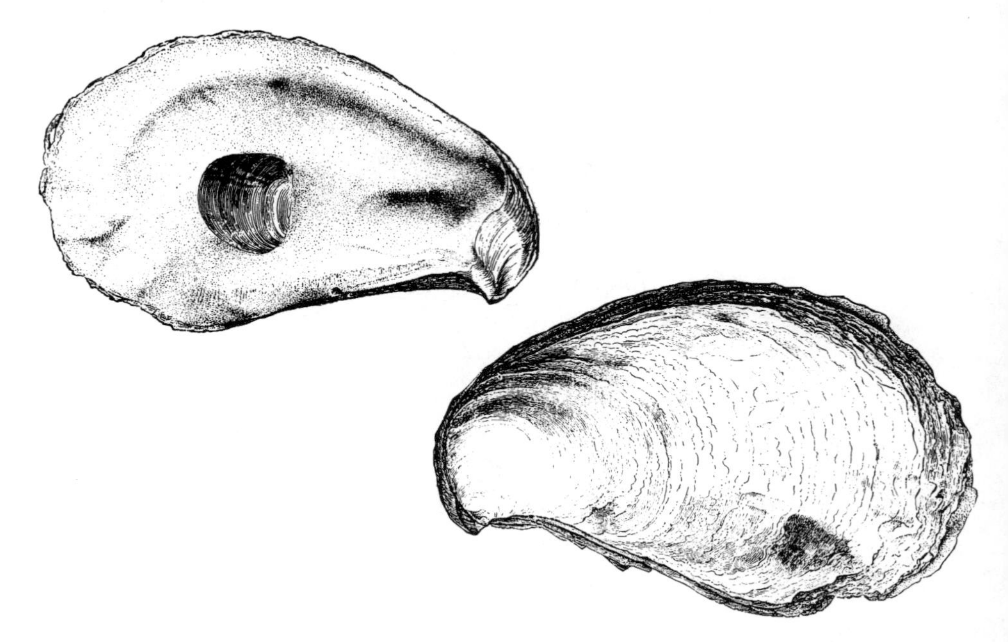

SCIENTIFIC AND COMMON NAMES: ***Crassostrea virginica,* American oyster**
Also called Eastern oyster.

DESCRIPTION AND DISTINCTIVE CHARACTERISTICS: The valve closest to the substrate to which the oyster is attached is cup-shaped; the upper shell is generally flat. Shell shapes vary greatly.

HABITAT RANGE: The American oyster is native to the Atlantic Coast of North America from the Canadian Maritimes southward around Florida, and all along the Gulf of Mexico to the West Indies and Venezuela. It has been introduced along the Pacific Coast in California and Washington, and also in Hawaii.

REPRODUCTION AND DEVELOPMENT: Spawning takes place when the temperature increases to within a range of 68–90 °F. Individual females release 14–114 million eggs free in the water; fertilization is external. After hatching, young drift as *veligers* (i.e., larvae equipped with a swimming organ, or *velum*) for two to three weeks; upon locating suitable substrate, they settle to the bottom, attach themselves, and remain there for their adult life.

AGE AND GROWTH: In the Maritime Provinces of Canada, 4–7 years to reach 3 in.; in Long Island Sound (U.S.), 1 year to reach 0.75 in., 3 to reach 3 in., and perhaps 20 to reach 15 in.; in the Gulf of Mexico, 2 years to reach 3.5 in.

FOOD AND FEEDING: Planktonic plants and animals are drawn into gills by the beating of many small hairs (*cilia*), then moved along by the hairs to the lips (labial palps) and mouth.

PARASITES AND DISEASES: A herpeslike virus, bacteria (*Vibrio* and unidentified), protozoans, and trematodes.

PREDATORS AND COMPETITORS: Predators include gastropods (oyster drills, whelks), starfish, flatworms, crabs, and fish; oysters are most vulnerable to predation when small and thin-shelled. Competitors include boring sponges, clams, mud worms, oyster crabs, mussels, tunicates, sponges, hydroids, bryozoans, ascidians, and algae.

AQUACULTURAL POTENTIAL: American oysters are well-suited for farming because they are in demand, grow rapidly, can be spawned artificially without great difficulty, and need not be impounded because they attach to the bottom. By using off-the-bottom rearing techniques, greater production of well-shaped oysters can be achieved. Extensive literature on species biology and results of aquacultural trials are readily available in government "how to" manuals.

REGIONS WHERE FARMED AND/OR RESEARCHED: U.S. East Coast and the Gulf of Mexico.

REFERENCES

Breisch, L. L., and V. S. Kennedy. 1980. *A selected bibliography of worldwide oyster literature*. University of Maryland Sea Grant publication UM-SG-TS-80-11. College Park: University of Maryland Sea Grant Program.

Burrell, V. B. 1985. Oyster culture. In *Crustacean and mollusk aquaculture in the United States*, ed. J. V. Huner and E. E. Brown, pp. 235–73. Westport, CT: AVI Publishing Co.

Galtsoff, P. S. 1964. The American oyster *Crassostrea virginica* Gmelin. *Fishery Bulletin* 64:1–480.

Shaw, W. N. 1969. The past and present status of off-bottom oyster culture in North America. *Transactions of the American Fisheries Society* 98:755–61.

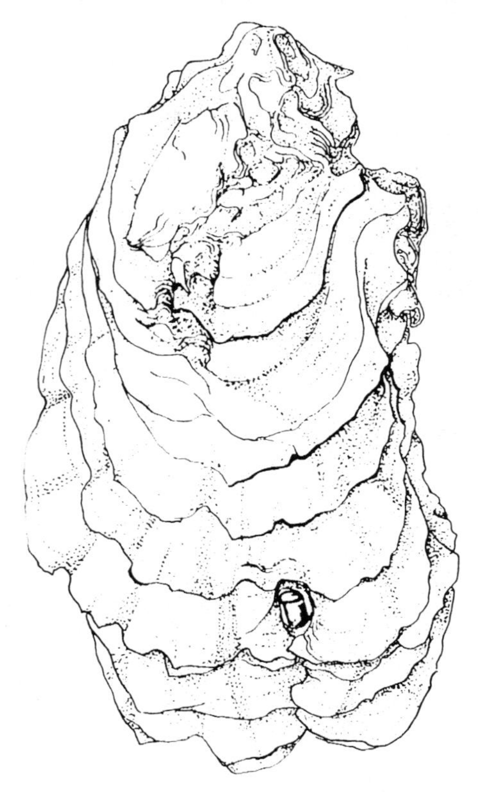

SCIENTIFIC AND COMMON NAMES: ***Crassostrea rhizophorae*, Mangrove oyster**

DESCRIPTION AND DISTINCTIVE CHARACTERISTICS: A small oyster that forms large clusters on red mangrove roots.

HABITAT RANGE: Southernmost Florida, Cuba, and Puerto Rico southward to Brazil; in waters with a good tidal exchange and salinities of 28–36 ppm.

REPRODUCTION AND DEVELOPMENT: Spawns January–May, releasing about 170 million eggs; young larvae drift for 20–60 days before settling on clean substrate (usually mangrove roots). Sexes are separate and/or hermaphroditic.

AGE AND GROWTH: Length usually about 2–3 in.; may reach 6 in.

FOOD AND FEEDING: Filter feeder.

PARASITES AND DISEASE: Nematode cysts and *Nematopsis* (protozoan) parasitize mangrove oysters.

PREDATORS AND COMPETITORS: Fish are important predators. Competitors include ascidians, sponges, barnacles, bryozoans, hydrozoans, and a triton (gastropod).

AQUACULTURAL POTENTIAL: The first commercial oyster farm in Cuba, located on the northeastern shore, began operations in 1975; about 20 farms were in operation in Cuba recently. Jamaica investigated the possibility of commercial mangrove oyster culture for about four years and began serious culture in 1980 on racks and rafts. Hatcheries are not important sources of spat, which is generally collected in the wild. Oyster production is increased by hanging twigs and branches for the larval oysters to settle on. Under certain conditions, mangrove oysters may grow faster if held subtidally rather than intertidally because continuous immersion permits continuous feeding; however, this increase in production must be balanced against the higher predation rates usually associated with such long-term submersion. Rapid growth and fine flavor, plus the reduction of natural habitats for this traditional seafood in many developing countries, make the mangrove oyster a good aquacultural candidate.

REGIONS WHERE FARMED AND/OR RESEARCHED: Cuba, Jamaica, Dominican Republic, Nicaragua, Venezuela, Brazil, Colombia, and Mexico.

REFERENCES

Jory, D. E., and E. S. Iversen. 1985. Molluscan mariculture in the greater Caribbean: An overview. *Marine Fisheries Review* 47(4):1–10.

Littlewood, D. T. J. 1988. A bibliography of literature on the mangrove oyster *Crassostrea rhizophorae* (Guilding, 1828). *Journal of Shellfish Research* 7(3):389–93.

Littlewood, D. T. J. 1991. Oyster cultivation in the Caribbean with an emphasis on mangrove oysters in Jamaica. *World Aquaculture* 22(1):70–7.

Pen Shell

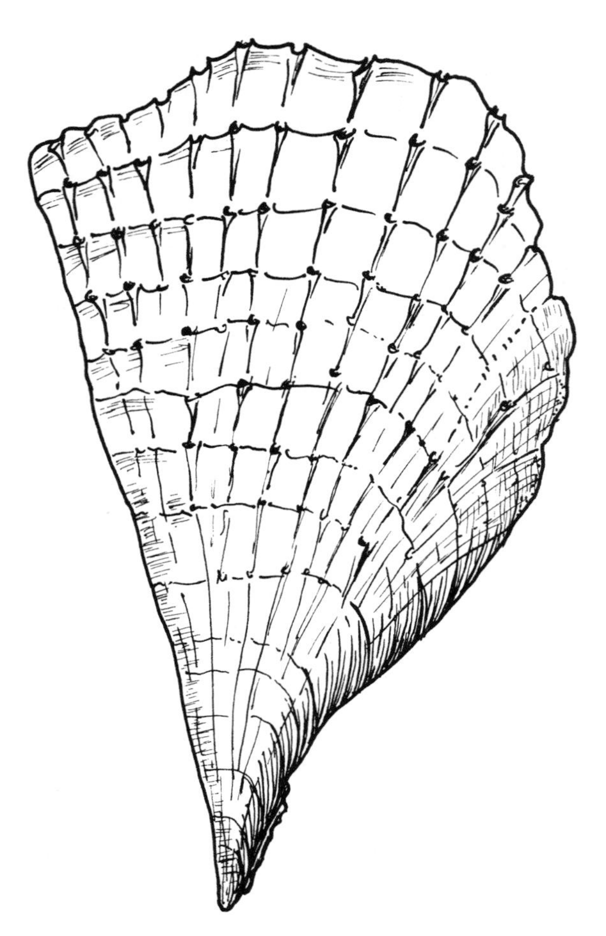

SCIENTIFIC AND COMMON NAMES: ***Pinna carnea, Atrina rigida***
Pen shell

DESCRIPTION AND DISTINCTIVE CHARACTERISTICS: Pen shells have a large, rounded-triangular, fragile shell, which seems oversized for the fleshy part of the animal it encloses. Byssus threads on the pointed end serve to anchor it in place. Numerous ridges and "teeth" are present on the outside of the shell. Shell coloration may vary from amber to light orange.

HABITAT RANGE: Found on sandy–muddy bottoms in which they bury the pointed end in the substrate; the large, blunt end of the shell is exposed.

REPRODUCTION AND DEVELOPMENT: Not reported.

AGE AND GROWTH: Spat collected during September in the Gulf of California; could reach a length of 10 in. within one year under good growth conditions.

FOOD AND FEEDING: Filter feeder.

PARASITES AND DISEASES: None reported.

PREDATORS AND COMPETITORS: Pea crab (commensal).

AQUACULTURAL POTENTIAL: Development may involve a long planktonic stage, which would make larval rearing difficult. On the plus side, pen shells seem to invest little energy developing viscera and shell, and therefore might grow to market size rapidly. Only the single large muscle is eaten. The opinion of gourmets is that it is difficult to find a sweeter and tastier shellfish in the sea. An added bonus for the pen shell lover is the possibility, if one is very lucky, that during a lifetime of eating pen shells, one may find a valuable pearl. Development of pen shell aquaculture has been stalled because of the lack of important scientific knowledge of the species; also, in some areas, wild pen shells have been rather abundant and free for the taking. Heavy fishing on natural stocks may increase the demand and prices so that aquacultural enterprises may be feasible.

REGIONS WHERE FARMED AND/OR RESEARCHED: Shellfish hatcheries in the Gulf of California (in the Mexican states Baja California Sur, Sonora, Sinaloa, and Nayarit) have experimented with culture techniques for pen shells.

REFERENCES

Baqueiro Cardenas, E. 1984. Status of molluscan aquaculture on the Pacific coast of Mexico. *Aquaculture* 39(1–4): 83–93.

Berg, C. J. 1984. Aquaculture potential of Bermudian bivalve mollusks. In *Assessment of the potential for aquaculture in Bermuda,* ed. T. D. Skeeter, pp. 77–89. Bermuda Biological Station Special Publication no. 27. Ferry Reach: Bermuda Biological Station.

Jory, D. E., and E. S. Iversen. 1985. Molluscan mariculture in the greater Caribbean: An overview. *Marine Fisheries Review* 47(4):1–10.

Scallops

Scallops are bivalves with two rather similar shells (*valves*) hinged by a very large adductor muscle; they are able to swim short distances by contracting this muscle and clapping their shells together. The two valves, described as "upper" and "lower," are actually left and right valves; the left or lower valve is usually flatter. Most species have a series of brightly colored eyes situated along their mantle's edge. Scallops in the United States inhabit nearshore and estuarine waters. After a short drifting life, the recently hatched young settle to the bottom.

The sea scallop (*Patinopecten yessoensis*) is successfully raised in Japan. In the United States and Canada, the bay scallop (*Argopecten irradians*) has undergone trials for aquacultural potential. *A. circularis* is raised in Baja California, Mexico, and *A. purpuratus* in Peru.

Because scallops are mobile and subject to predation, they are usually raised in enclosures (e.g. nets hung vertically in the water column). Automatic shucking devices are available for some scallop species; these could be modified to process other scallop species, thus alleviating labor and other problems related to hand shucking operations.

Although there is definite interest in scallop culture in North America, none of the trials made in California, Washington, Alaska, or British Columbia to produce dependable supplies of seed has been encouraging; but after about 20 years of aquacultural R&D trials on the Pacific Coast of Canada using the bay scallop and three other species—the Japanese (sea) scallop on the Canadian Pacific Coast, and pink and weathervane scallops in Puget Sound and the Strait of Georgia (using lantern nets)—most of the early biological constraints seem to have been overcome. Nevertheless, high production costs (bottom culture seems to be the only economic method) and stiff competition from low-priced scallops from capture fisheries and foreign imports are obvious hurdles that must be cleared before Canadian scallop farming can be successful. With the success of scallop culture in Japan as a stimulus, however, plus strong Pacific Coast markets and numerous areas suitable for culture, such an industry may yet become a reality.

The rock scallop, also called the purple-hinge scallop (*Crassadoma gigantea* = *Hinnites multirugosus*), attaches itself to a clean, firm surface and has been suggested as a candidate for aquaculture. Somewhat limited biological information exists for this species at the present time, and only a single culture facility (in California) raises the rock scallop as a second crop.

REFERENCES

Bourne, N. 1991. West coast of North America. In *Scallops: Biology, ecology and aquaculture*, ed. S. E. Shumway, pp. 925–42. Developments in Aquaculture and Fisheries Science, vol. 21. Amsterdam: Elsevier.

Bourne, N., C. A. Hodgson, and J. N. C. Whyte. 1989. *A manual for scallop culture in British Columbia.* Canadian Technical Report of Fisheries and Aquatic Sciences no. 1694. Nanaimo, BC: Fisheries and Oceans Canada.

Costello, T. 1985. Peru has thriving scallop industry. *Aquaculture Magazine* 11(3):32-36.

Couturier, C. 1990. Scallop aquaculture in Canada: Fact or fantasy? *World Aquaculture* 21(2): 54–62.

Shumway, S. E., ed. 1991. *Scallops: Biology, ecology and aquaculture.* Developments in Aquaculture and Fisheries Science, vol. 21. Amsterdam: Elsevier.

Winter, J. E., J. E. Toro, J. M. Navarro, G. S. Valenzuela, and O. R. Chaparro. 1984. Recent developments, status and prospects of molluscan aquaculture on the Pacific Coast of South America. *Aquaculture* 39(1–4):95–134.

SCIENTIFIC AND COMMON NAMES: ***Argopecten irradians,* Bay scallop**

DESCRIPTION AND DISTINCTIVE CHARACTERISTICS: Upper valve is marked with mottling of reds, tans, yellows, and purples. Both valves are inflated.

HABITAT RANGE: Nearshore and estuarine waters along the Atlantic Coast of the United States, from Cape Cod to Texas.

REPRODUCTION AND DEVELOPMENT: Spawn at about 81–86 °F when about 1 year old; unlike most scallops, this species is *hermaphroditic* (has both sexes in the same individual). Larvae begin feeding on algae after 2–3 days; after about 5–8 days, they settle to the bottom, attach themselves to vertical surfaces by threads, and transform into juveniles.

AGE AND GROWTH: In the wild, bay scallops reach market size (2–2.5 in.) in about 16 months, and have a life span of roughly two years.

FOOD AND FEEDING: Scallops feed by filtering algae, diatoms, protozoans, and detritus. Occasionally crustaceans, polychaetes, echinoderms, and seaweeds are found in their stomachs.

PARASITES AND DISEASE: Bacteria, flatworms (flukes), nematodes (roundworms), mollusks, and arthropods (crustaceans).

PREDATORS AND COMPETITORS: The starfish is one of the major scallop predators, either eating the animal whole or inserting its stomach between the shells it pulls open; snails, crabs, fish, and water birds also feed on these scallops.

AQUACULTURAL POTENTIAL: Short larval, juvenile, and adult stages, high market value and price, established hatchery techniques, and low feed costs are all strong reasons for this species to become important in aquaculture. Private shellfish hatcheries sell bay scallop seed. The bay scallop has undergone trials for aquacultural potential on the U.S. Atlantic Coast. The economics of U.S. scallop farming and locations suitable for scallop culture require careful scrutiny before the potential of commercial ventures can be evaluated.

REGIONS WHERE FARMED AND/OR RESEARCHED: Connecticut, Georgia; Canada.

REFERENCES

Couturier, C. 1990. Scallop aquaculture in Canada: Fact or fantasy? *World Aquaculture* 21(2): 54–62.

Heffernan, P. B., R. L. Walker, and D. M. Gillespie. 1988. Biological feasibility of growing the northern bay scallop, *Argopecten irradians irradians* (Lamarck, 1819) in coastal waters of Georgia. *Journal of Shellfish Research* 7(1):83–8.

Leighton, D. L. 1991. Culture of *Hinnites* and related scallops on the American Pacific coast. In *Estuarine and marine bivalve mollusk culture*, ed. W. Menzel, pp. 99–111. Boca Raton, FL: CRC Press.

Middleton, K. C. 1983. Bay scallops: A mariculture species whose time has come. *Aquaculture Magazine* 9(6):16–20.

Rhodes, E. W. 1991. Fisheries and aquaculture of the bay scallop, *Argopecten irradians*, in the eastern United States. In *Scallops: Biology, ecology and aquaculture*, ed. S. E. Shumway, pp. 913–24. Developments in Aquaculture and Fisheries Science, vol. 21. Amsterdam: Elsevier.

Gastropods

Mollusks in the class Gastropoda, such as conchs (genus *Strombus*) and whelks (genus *Busycon*), have a single, coiled shell with a horny operculum that seals the shell opening when danger approaches. These marine snails feed by rasping their food with a ribbonlike series of teeth called *radulae*. Their diet consists, generally, of plant material and possibly some benthic organisms. They move moderate distances by means of their foot, which, when extended, either lets them slide or glide over the bottom on a path of excreted slime, or pushes them along with a sort of hopping motion. Their geographic range is wide, with different species found in limited areas within the range.

Whelks extend from Cape Cod to Florida; though formerly considered pests by oyster farmers and fishers, they have recently come to have significant economic importance in the New England area. However, the gastropods with the greatest aquacultural potential are abalones, queen conchs, and top shells.

REFERENCE

Hahn, K. O., ed. 1988. *Handbook of culture of abalone and other marine gastropods.* Boca Raton, FL: CRC Press.

Abalone

SCIENTIFIC AND COMMON NAMES: ***Haliotis rufescens,* Red abalone**

DESCRIPTION AND DISTINCTIVE CHARACTERISTICS: The abalone has a large, powerful, flat, muscular foot with which it attaches to rocks or creeps over the bottom. The large shell has holes over the gill cavity. Sizable ones measure over 11 in. and weigh over 4 lb.

HABITAT RANGE: Sitka, Alaska, to Baja, California. Intertidal and sublittoral; found on rocks, under ledges, and in fissures where there is an abundance of drift algae and vegetation.

REPRODUCTION: Eggs released free in the sea sink to the bottom; planktonic larvae settle to the bottom after 5–14 days. Fecundity may be as high as 100,000 to 2.5 million eggs per day.

AGE AND GROWTH: Slow growth (1.13 in./yr) may require holding abalones for 2.5–3 years until ready for market.

FOOD AND FEEDING: Young eat diatoms, unicellular algae, and occasionally coralline algae. Juveniles and adults eat macroalgae such as *Ulva* and *Macrocystis,* although different species prefer different algae. Grazing on the bottom, abalones eat small animals such as hydrozoons, copepods, foraminifera, and byrozoans.

PARASITES AND DISEASE: A nematode, *Echinocephalus pseudouncinatus,* has been reported from the pink abalone. Bacteria (*Vibrio alginolyticus*) was found in cultured abalone.

PREDATORS AND COMPETITORS: Predators include rockfish, sea otters, starfish, octopus, whelks, rock crabs, and rock lobsters.

AQUACULTURAL POTENTIAL: Abalones are commercially important around Hawaii and along the U.S. Pacific Coast; their rather low potential is based on slow growth rates. Natural mortality of unprotected young abalone is very high. Strong market due to desirable taste and overfishing by commercial and recreational fisheries.

REGIONS WHERE FARMED AND/OR RESEARCHED: California, Hawaii; Korea, Ireland, Japan, and Australia.

REFERENCES

Ault, J. S. 1985. *Species profiles: Life histories and environmental requirements of coastal fishes and invertebrates (Pacific Southwest). Black, green, and red abalones.* U.S. Fish and Wildlife Service Biological Report 82(11.32); U.S. Army Corps of Engineers TR EL-82-4. Slidell, LA: U.S. Fish and Wildlife Service.

Hahn, K. O., ed. 1988. *Handbook of culture of abalone and other marine gastropods.* Boca Raton, FL: CRC Press.

Johnson, T. 1989. Farming sea-ears; an abalone success story. *Sea Frontiers* 35(4):232–7.

Nash, C. E. 1991. The production of abalone. In *Production of aquatic animals; crustaceans, molluscs, amphibians and reptiles,* ed. C. E. Nash, pp. 173–81. Amsterdam: Elsevier.

Conch

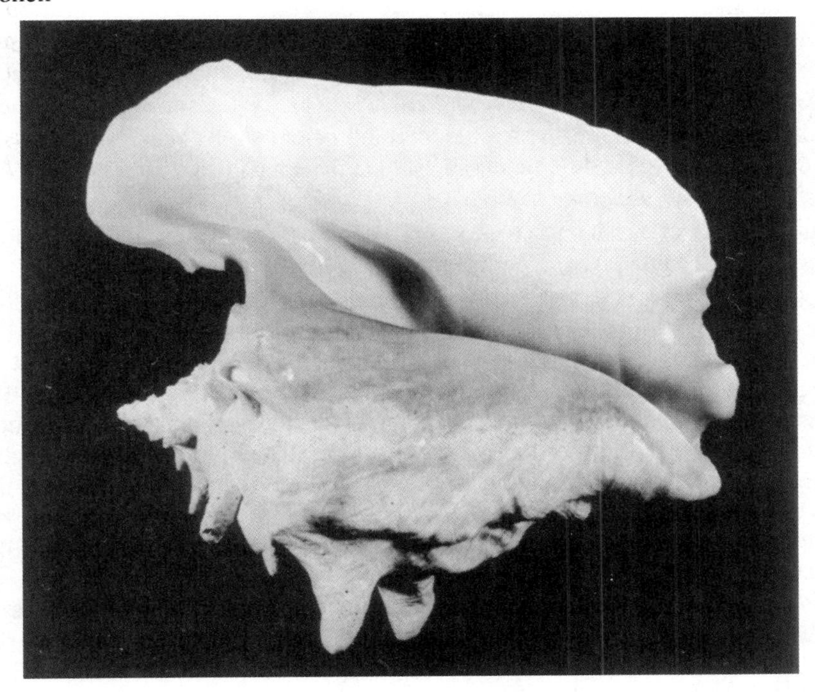

SCIENTIFIC AND COMMON NAMES: ***Strombus gigas,*** **Queen conch** Related: milk conch, *S. costatus; S. galeatus* (Pacific); *S. goliath* (Brazil).

DESCRIPTION AND DISTINCTIVE CHARACTERISTICS: Large (about 1 ft long) colorful marine snail. Can move short distances on a heavy muscular foot, not by gliding on a path of slime like some gastropods, but rather by extending its foot, raising the shell off the bottom, and pushing itself forward.

HABITAT RANGE: Southeastern Florida, southern Gulf of Mexico, the Caribbean Sea, and the Bahamas and Bermuda; from the intertidal zone to 200 m depth.

REPRODUCTION AND DEVELOPMENT: Queen conch usually become sexually mature at the "flared lip" stage. Females spawn several times during the summer, producing large egg masses containing hundreds of thousands of eggs per spawning; these masses lie on the seafloor. Upon hatching, the young *veligers* (swimming larvae) drift in the plankton for about three weeks before settling and taking up a bottom-dwelling existence. Like other mollusks, they may use cues to determine when to settle, beginning a metamorphosis from planktonic to bottom-dwelling forms; this may require 20 hours to complete. When the planktonic stages become competent to settle, they must do so in 3–4 days.

AGE AND GROWTH: Queen conchs reach market size (about 1 ft) in roughly 3–3.5 years.

FOOD AND FEEDING: Veligers consume algae; upon settling to the bottom, juvenile and adults eat detritus and microalgae. Its *crystalline style* (translucent rod) releases digestive enzymes (microprotein gel) to aid in digesting carbohydrates.

PARASITES AND DISEASES: An unidentified nematode has been found in the ganglion of this conch. Commensals include the conch fish, *Astrapogon stellatus,* and a crab, *Porcellana sayana.*

PREDATORS AND COMPETITORS: Small octopuses, hermit crabs, spiny lobsters, spotted eagle rays, loggerhead turtles, tulip shells, and finfishes.

AQUACULTURAL POTENTIAL: Queen conch meat is a high-value product in Florida and the Caribbean Islands, as well as a basic subsistence food; it has a fine clamlike flavor and is used in chowder, salad, or fritters. The conch shell yields other products: ornaments or souvenirs for tourists, a type of "pearl" that occurs very infrequently in wild conchs, and tools for aboriginal peoples; moreover, shells are burned for lime to make mortar for buildings. (There are also unsubstantiated claims that conchs possess antitumoral properties, and that the crystalline style is an aphrodisiac.) Yet despite the advantages of a high market demand and its feeding low on the food web, the rather slow growth of the queen conch does not make it a very desirable species for aquaculture.

REGIONS WHERE FARMED AND/OR RESEARCHED: DNR Field Station, Marathon, Florida; University of Miami, RSMAS and Undergraduate Marine Science Department; Cancún, Quintana Roo, Mexico; Los Roques, Venezuela; Belize; Instituto de Oceanologia, Havana, Cuba; NOAA Underwater Research, St. Croix, Virgin Islands; University of Puerto Rico, Mayagüez; Martinique; Caicos Conch Farm, Providenciales, Turks and Cacios Islands; Division of Fisheries, Bermuda; St. Lucia and St. Kitts–Nevis, British West Indies; Caribbean Research Center, Vero Beach, Florida (field station = Lee Stocking Island, Exuma Cays, Bahamas).

REFERENCES

Brownell, W. N. 1977. Reproduction, laboratory culture, and growth of *Strombus gigas, S. costatus,* and *S. pugilis* in Los Roques, Venezuela. *Bulletin of Marine Science* 27:668–80.

Brownell, W. N., and J. M. Stevely. 1981. The biology, fisheries and management of the queen conch, *Strombus gigas. Marine Fisheries Review* 43(7):1–12.

Darcy, G. H. 1981. *Annotated bibliography of the conch genus,* Strombus *(Gastropoda, Strombidae) in the western Atlantic Ocean.* NOAA Technical Report NMFS SSRF-748. Seattle: National Marine Fisheries Service.

Hahn, K. O., ed. 1988. *Handbook of culture of abalone and other marine gastropods.* Boca Raton, FL: CRC Press.

Top Shell

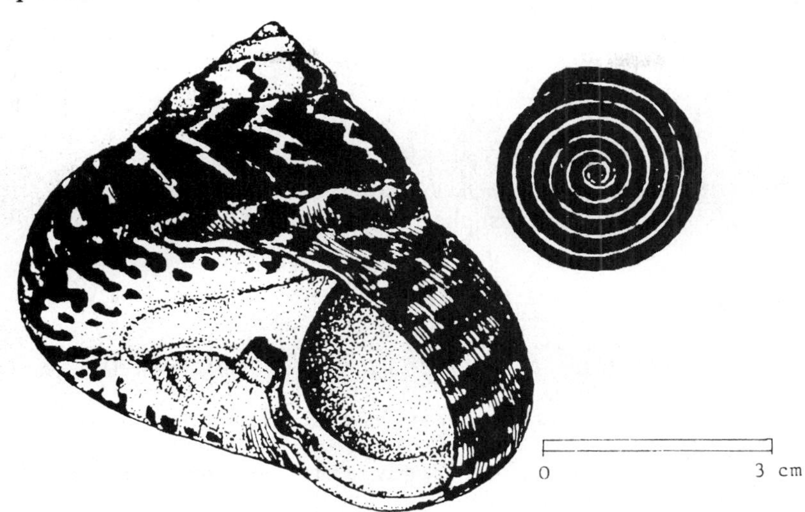

SCIENTIFIC AND COMMON NAMES: ***Cittarium (Livona) pica,* West Indian top shell**
Related: Pacific top shell, *Trochus niloticus*

DESCRIPTION AND DISTINCTIVE CHARACTERISTICS: A heavy-shelled gastropod, shaped like a top with numerous whorls and a sharp outer lip protected by a horny *operculum* (lidlike structure). Exterior is grayish cream with zigzagging purplish-black blotches.

HABITAT RANGE: In close contact with seashores, from high tide mark down to depths of about 3 ft. Usually found in crevices on rocky or coral areas with heavy surf. Extending from the Bahamas and Yucatán, Mexico, southward to Trinidad.

REPRODUCTION AND DEVELOPMENT: Spawning takes place year round, with greatest intensity during the winter.

AGE AND GROWTH: Reaches maximum "length" of about 5.25 in. at age 5–6 years.

FOOD AND FEEDING: Herbivorous; grazes on algae (blue-green, green, red, and brown) as well as on diatoms and organic detritus.

PARASITES AND DISEASE: No reports.

PREDATORS AND COMPETITORS: Eaten by rock shells (*Purpura* spp. and *Thais* spp.), a variety of finfishes, and some birds (which occasionally eat the smaller top shells). A small limpet and a pinnotherid crab are commensals with the top shell.

AQUACULTURAL POTENTIAL: Widely used as human food in the West Indies and Bahamas; some locals prefer top shell over the large queen conch. Although suggested for artificial culture—there are no large feed costs, very little labor, and apparently no known disease or parasite problems (at least in the juveniles and adults), and local popularity is high—no serious trials have been reported in the literature. The Pacific species, *T. noliticus,* is marketed in Japan and Europe for the meat and the shells (for buttons).

REGIONS WHERE FARMED AND/OR RESEARCHED: (*T. niloticus* only) Torres Strait (off northeastern tip of Australia); Palau, Caroline Islands, Micronesia.

REFERENCES

Debrot, A. O. 1987. *Ecology and management of the West Indian top shell* Cittarium pica *(L.) (Gastropoda : Trochidae) of the Exuma Islands, Bahamas.* Ph.D. dissertation. Coral Gables, FL: University of Miami.

Hahn, K. O., ed. 1988. *Handbook of culture of abalone and other marine gastropods.* Boca Raton, FL: CRC Press.

Heslinga, G. A., and A. Hillmann. 1981. Hatchery culture of the commercial top snail *Trochus niloticus* in Palau, Caroline Islands. *Aquaculture* 22:35–43.

Randall, H. A. 1964. A study of the growth and other aspects of the biology of the West Indian topshell, *Cittarium pica* (Linnaeus). *Bulletin of Marine Science of the Gulf and Caribbean* 14:424–43.

VERTEBRATES

Vertebrates are animals that have a vertebral column (backbone). They comprise a large division (subphylum Vertebrata) of the cordates (phylum Chordata), and include five classes: fishes, amphibians, reptiles, birds, and mammals. Aquacultural products fall into only the first three of these classes; however, mammals and especially birds may be important predators on cultured species.

Amphibians

Animals in this class of vertebrates spend most or all of their lives in water or damp places. Most of them lay eggs in the water, producing larvae that breathe with gills (i.e., tadpoles or pollywogs); after metamorphosis, adults breathe with lungs. Amphibians are cold-blooded and have a three-chambered heart. Their skins are smooth, slimy, and (with few exceptions) scaleless. The only group of interest to aquaculturists is frogs, for whose meat and skin there is a good market.

REFERENCE

Culley, D. D. 1986. Bullfrog culture still a high risk venture. *Aquaculture Magazine* 12(5): 28–35.

Frog

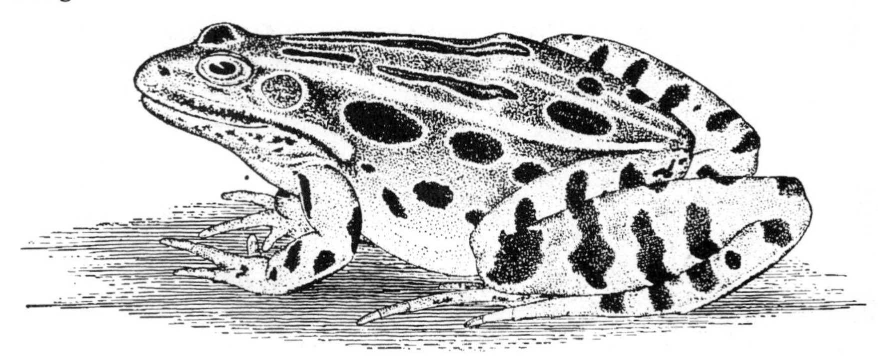

SCIENTIFIC AND COMMON NAMES: ***Rana catesbeiana,* American bullfrog**
There are about 20 species of frogs, but the American bullfrog is used most often for culture.

DESCRIPTION AND DISTINCTIVE CHARACTERISTICS: Large freshwater amphibian with thin, moist skin. Their forelegs are short; hind legs are powerful with webbed feet. They are tailless as adults and well adapted for leaping.

HABITAT RANGE: Native to U.S. South and northern Mexico. Widely transplanted to Canada, Europe, Cuba, Asia, Latin America, and many Pacific Ocean islands.

REPRODUCTION AND DEVELOPMENT: Eggs are laid in water and hatch into tadpole (gilled, larval) stage, which later metamorphoses into adult form. A large female may lay 20–25 thousand eggs at one time, and may spawn as many as 10–12 times per year.

AGE AND GROWTH: Approximately 6–10 months are required to reach market size (200–250 g). A record weight and length of 7.5 lb and 2 ft has been reported from Africa. They may live as long as 16 years.

FOOD AND FEEDING: Eat live crickets, earthworms, insects, and minnows. In culture, fly larvae (maggots) may be raised and fed to frogs. Under certain conditions, frogs can be cannibalistic.

PARASITES AND DISEASE: Bacterial diseases are important in culture; at least 15 pathogenic bacteria are recognized.

PREDATORS AND COMPETITORS: Fish and birds are the major predators, although frogs are eaten by a wide variety of aquatic and terrestrial vertebrates, including alligators, aquatic turtles, snakes, raccoons, and otters. Large individuals will cannibalize small ones.

AQUACULTURAL POTENTIAL: Good market, though the human food market is not as good as that for research institutions. The skin, which is sold for leather, is the most promising market product. Frogs can be raised quickly and in large numbers; however, expenses are high, and it is difficult to compete costwise with frogs that appear naturally in fish culture facilities (and grow to market size with no care or expense to the fish farmers) or those caught in the wild. Also, the reluctance of frogs to take nonliving food has been a deterrent to their farming. Exports from Brazil and Asia can reduce the market price of U.S.-grown frogs. High density of frogs results in cannibalism and low weight gain per individual. Outdoor rearing is considered obsolete because of high predation, disease, and cannibalism.

REGIONS WHERE FARMED AND/OR RESEARCHED: California and Brazil.

REFERENCES

Adams, I. K., and A. C. Bruinsma. 1987. Intensive commercial bullfrog culture: A Brazilian experience. *Aquaculture Magazine* 13(4):28–44.

Culley, D. D. 1986. Bullfrog culture: Still a high risk venture. *Aquaculture Magazine* 12(5): 28–35.

Culley, D. D. 1991. Bullfrog culture. In *Production of aquatic animals; crustaceans, molluscs, amphibians and reptiles,* ed. C. E. Nash, pp. 185–205. Amsterdam: Elsevier.

Culley, D. D., and C. T. Gravois. 1970. Frog culture: A new look at an old problem. *American Fish Farmer* 1(5):5–10.

Lester, D. 1988. Raising bullfrogs on non-living food. *Aquaculture Magazine* 14(2):20–7.

Priddy, J. M., and D. D. Culley. 1972. Frog culture industry; past, present, future? *American Fish Farmer & World Aquaculture News* 3(9):4–7.

Finfishes

Finfishes (class Pisces) are cold-blooded, lower aquatic vertebrates possessing fins and (usually) scales. They are also called "true fishes" to differentiate them from other organisms that have the word "fish" in their names (e.g., crayfish, starfish). Most fishes with aquacultural applications are *teleosts,* or fishes with bony skeletons and hinged jaws.

Some Aquaculturally Important Fish Groups

Common names are often used to refer to more than one species of fish, and can even refer to fishes from different families: The seatrout, for example, is not a trout (Family: Salmonidae) but a drum (Family: Sciaenidae). Scientific classification of fishes into families, genera, and species is intended to alleviate such problems by assigning one name to each fish and organizing them hierarchically. This system may appear overtechnical and confusing to the layperson, and at times equally confusing to the scientist: Some fishes, such as the rainbow trout (steelhead), have been reclassified, their genus and species completely changed. Still, since family or other group names do crop up in the literature, several of the more aquaculturally inclusive are given overviews below. (Discussions of individual species in these and other groups are organized *alphabetically by keyword* [largemouth *bass, buffalo* fish, Arctic *char,* red *drum,* etc.] in the later section "Aquaculturally Important Fish Species.")

Centrarchids

(*See:* bass, crappie, pumpkinseed)

The family Centrarchidae, commonly called the sunfish family, comprises numerous species, including small- and largemouth bass, sunfish, bream, bluegills, pumpkinseeds, and crappies; several are raised for intensive and/or extensive aquaculture. (The common names sunfish, bream, and bluegill each refer to several species in the genus *Lepomis,* and hence are virtually useless for correct species identification.) All centrarchids are freshwater fish, generally found in warm to temperate lakes with abundant vegetation, and range from southern Canada to the Gulf of Mexico. They are considered very good panfish.

Centrarchids have rather deep bodies flattened from side to side, with two dorsal fins in the middle of the back (the front one spiny, the posterior one made of soft rays). They are nest builders, with the male building the nest and caring for the young. The number of eggs produced by a single female varies greatly from species to species and with the size of the female within each species: Egg counts as high as 50,000 have been made in a single bluegill nest, and 67,000 eggs were found in a single female.

The sunfish family played a major role in early U.S. warm-water pond culture programs. Many farms had ponds to supply water for livestock, crop irri-

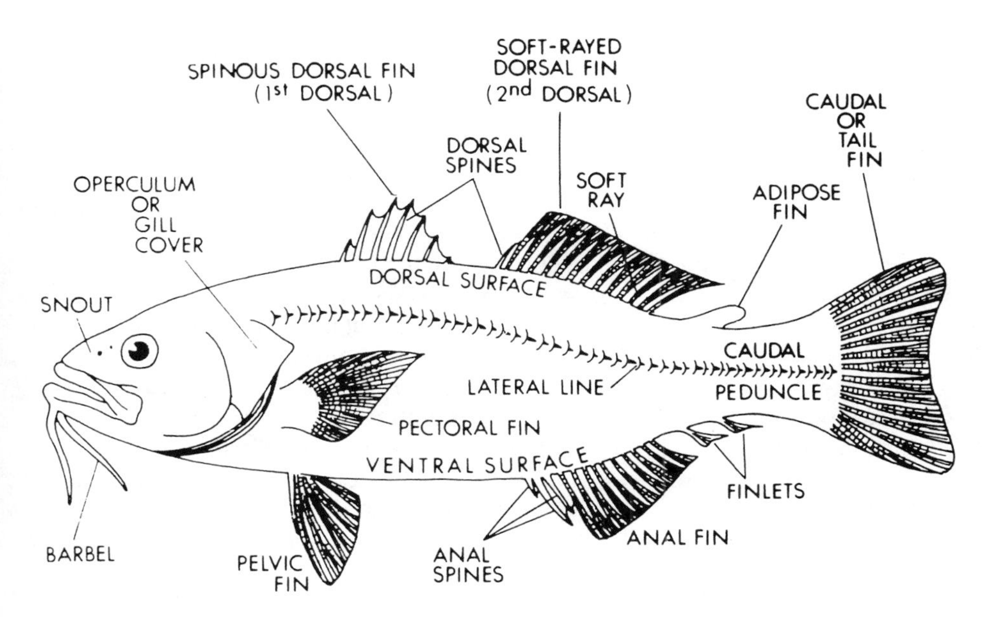

Hypothetical bony fish.

gation, and so on. In the 1930s the desire for sportfish and the need to conserve water and soil resulted in a Farm Pond Program begun by the U.S. Department of Agriculture. This program made a substantial contribution in providing technical assistance for the planning and construction of fish ponds, and federal hatcheries provided fingerlings to stock these ponds. Numerous rearing trials with centrarchids provided critical data on stocking practices. (See the section in Chapter 2 on private U.S. farming.)

In an effort to improve the aquacultural value of some of the sunfish, various species have been crossed. (It had been known that different species hybridized in nature.) The best hybrid for rapid growth, good condition, and high catchability for Texas waters was between the bluegill (*Lepomis macrochirus*) and the green sunfish (*L. cyanellus*). The green sunfish has been crossed with pumpkinseeds (*L. gibbosus*) to produce larger fish. A variety of crosses have been tried: the male bluegill with the female green, the male bluegill with the female redear (*L. microlophus*), and the male green with the female redear.

REFERENCES

McLarney, W. 1984. *The freshwater aquaculture book; a handbook for small scale fish culture in North America.* Point Roberts, WA: Hartley & Marks.

Stickney, R. R., ed. 1986. *Culture of nonsalmonid freshwater fishes.* Boca Raton, FL: CRC Press.

Swingle, H. S. 1970. History of warmwater pond culture in the United States. In *A century of fisheries in North America*, ed. N. G. Benson, pp. 95–105. American Fisheries Society Special Publication no.7. Washington, DC: American Fisheries Society.

Flatfish

(*See:* [Atlantic] halibut)

A most interesting fact about flatfish (Order: Heterosomata) is that they begin life swimming upright like most other fishes; but later they turn and lie on their sides (left or right, depending on the species), and one eye migrates across the forehead to the side that will be up. The skin pigment is lost on the side that will be in contact with the bottom; the top side remains pigmented and, in older fish, blends with the color of the bottom—another very interesting adaptation. Transformation from upright swimming to bottom-dwelling habits takes place in some flounder before they are half an inch long; once transformed they spend the remainder of their lives swimming on their sides, living on or close to the bottom. Some flatfish even bury themselves in muddy bottoms, leaving just their eyes exposed; then, when suitable prey species venture close, they dart out of the mud with surprising speed and capture the victim.

Characteristics used in identifying flatfishes are that some, including the turbot (Family: Bothidae), are "lefteye" (i.e., an eye migrates over the head to the left side). Bothidae and Soleidae (true soles) are generally circular in outline when viewed from above. Several species marketed as "sole" (e.g., rock sole, yellowfin sole, English sole) are not true soles but actually in the flatfish family Pleuronectidae. Flounder, dab, plaice, and halibut are also in this "righteye" family, and are thinner in outline than turbots and true soles. The largest commercial species of flatfish, and a very important one, is the halibut, which occurs in the Atlantic and Pacific oceans. Presently, only the Atlantic halibut is seriously considered for aquaculture.

Despite many years of experimental work in the United Kingdom beginning during World War II, attempts at profitable aquaculture of plaice (*Pleuronectes platessa*) have been unsuccessful. At first mortalities of 90 percent were caused by bacterial diseases in artificial cultures. This was overcome, but slow growth and rather expensive feed costs still remained serious problems to economic farming of the plaice. The sole (*Solea solea*) was also a candidate, but like the plaice was not mass-reared to market sizes on an economic basis. Reasons for failure of sole-rearing were essentially the same as those for the plaice.

The turbot (*Scophthalmus maximus*), another popular and highly valuable flatfish in the European market, seems to be the best flatfish for aquaculture tried so far. Fry production has been a difficult hurdle to overcome since the first attempts to culture the turbot artificially in Europe during the 1970s; but after almost 20 years of research in Scotland, Spain, France, Norway, and Denmark, there is considerable enthusiasm that commercial turbot farming will become successful.

REFERENCES

Cachelou, F., D. Duche, and A. Jones. 1989. Current status of turbot farming in Europe. *World Aquaculture* 20(3):50–3.

Cole, H. A. 1968. Fish rearing—will sole prove to be the best bet? *World Fishing* 17(4):26–7.

Ingram, M. 1987. The flatfish are coming. *Aquaculture Magazine* 13(3):44–7.

Person-Le Ruyet, J., F. Baudin-Laurencin, N. Devauchelle, R. Metailler, J. L. Nicolas, J. Robin, and J. Guillaume. 1991. Culture of turbot (*Scophthalmus maximus*). In *CRC handbook of mariculture, Vol. II: Finfish aquaculture*, pp. 21–41. Boca Raton, FL: CRC Press.

Riley, J. D., and G. T. Thacker. 1963. Marine fish culture in Britain. III. Plaice (*Pleuronectes platessa* L.) rearing in closed circulation at Lowestoft, 1961. *Journal du Conseil* 28(1):80–90.

White, D. B., and R. R. Stickney. 1973. *A manual of flatfish rearing*. University of Georgia Marine Science Center Technical Report Series no. 73-7. Savannah, GA: Skidaway Institute of Oceanography.

Salmonids

(*See:* [Arctic] char, [Atlantic] salmon, [chinook] salmon, [steelhead] trout)

The family Salmonidae includes both salmons and trouts, which are found in both the Atlantic and Pacific oceans. They have no spiny rays in their fins, only soft rays, and there is a small, fleshy adipose fin located between the dorsal and caudal fin. Highly migratory fishes, some travel fantastic distances but are still able to return home to the streams or rivers they left as small juveniles; there they build nests and spawn. The eggs are large and yolk-filled; the tiny larvae are born with a good supply of yolk to begin what for some species will be a long and dangerous journey.

Pacific salmons—such as the chinook (*Oncorhynchus shawytscha*), pink (*O. gorbuscha*), coho (*O. kisutch*), and steelhead (*O. mykiss* [formerly called rainbow trout, *Salmo gairdneri*])—stop feeding once they enter fresh water, where they spawn and die. (Exceptions are precocious male salmon and steelheads.) Some trout spend their entire lives in fresh water, and even the most migratory species of salmon may become landlocked and never leave fresh water.

Salmonids have been widely transplanted to many U.S. states and countries worldwide. Coho and pink salmons have been transplanted to the U.S. Great Lakes, and pink and chinook salmons have reproduced and become established there. Salmonids have been commercially profitable both in marine net pens (for Atlantic, chinook, and coho salmons) and in raceways on land (for trout). In Norway, with increased disease problems in high-density net pens, farmers have used antibiotics to protect their crops; however, concern has arisen that these antibiotics significantly threaten the environment by producing resistant bacteria. In Asia, the cherry salmon (*O. masou*) is used in large-scale rearing programs to augment wild stocks.

The literature available on all aspects of rearing salmonid fishes is voluminous. Quite popular as food fish, they have been spawned artificially and raised for release to augment natural stocks in U.S. fisheries since the mid-1800s; the large eggs make spawning and early rearing easy, but huge numbers of eggs must be obtained, hatched, and raised to a size at which they can be released in the wild and still avoid some of their enemies. Hatcheries that raise fish to enhance natural stocks are generally either public facilities or private, government-contracted agencies that supply certain quantities and sizes of fish for release programs. In these programs, meant to help the fishing industry, dollar profit is not normally the objective; the focus instead is on how many animals can be released, and how many of these are subsequently caught by the fisheries.

Ranching, on the other hand, is only about 20 years old, and has not yet been made profitable for private industry. Salmon ranchers—like their counterparts, cattle ranchers—release their stock to roam about to find food, thereby avoiding the usual high cost of feeds required for captive animals. The difference is that fish ranchers cannot saddle up and drive their salmon to market; they must wait for them to return, relying on the salmonid ability to home in on native streams. Year after year, extremely large numbers of healthy salmon eggs must be obtained and a huge, efficient hatchery system continue to raise and release young salmon—all while waiting for the adults to return several years later. Several important disadvantages are thus inherent in such operations: They require great cash investments and rely on high returns of salmon that have been raised at great expense and effort. Many new businesses may feature slow returns on investment, but the risk in salmon ranching is greater than usual. Enormous financial losses have been sustained by large-scale sea ranching operations in the Pacific Northwest (Oregon, Alaska) and Chile, due to very poor adult-salmon returns attributable to any number of causes: straying (failure to return to the home stream); catches by commercial and recreational fishers; high natural mortality (mostly due to predation by sea lions, seals, and salmon sharks); exceeding the carrying capacity of estuarine release sites; and so on.

There are a number of trout species, but only two true trouts are commonly used in aquaculture: the brown trout and the brook trout. The brown trout (*Salmo trutta*) is used for stock enhancement or "put and take" fisheries. The brook trout (*Salvelinus fontinalis*) is more popular with aquaculturists because it has greater disease resistance and grows faster than the brown trout. The steelhead or rainbow trout is also aquaculturally important; however, as mentioned earlier, it has recently been reclassified as a salmon.

REFERENCES

Edwards, D. J. 1978. *Salmon and trout farming in Norway.* Farnham (U.K.): Fishing News Books.

Laird, L. M., and T. Needham, eds. 1988. *Salmon and trout farming*. New York: John Wiley.
Sedgwick, S. D. 1988. *Salmon farming handbook*. Farnham (U.K.): Fishing News Books.
Svrjcek, R. S., ed. 1991. *Marine ranching; proceedings of the 17th U.S.–Japan Meeting on Aquaculture, Ise, Mie Prefecture, Japan, October 16–18, 1988*. NOAA Technical Report NMFS 102. Seattle, WA: National Marine Fisheries Service.
Thorpe, J. E., ed. 1980. *Salmon ranching*. New York: Academic Press.

Sciaenids

(*See:* [Atlantic] croaker, [red] drum, seatrout)

Of the thirty or more North American species in the large family Sciaenidae (croakers and drums), most are valued for their fine flesh. A few species, namely the Atlantic croaker (*Micropogonias undulatus*), the red drum (*Sciaenops ocellatus*), and the seatrout (*Cynoscion nebulosus*), have attracted the attention of aquaculturists. The body of such fishes is compressed and somewhat elongated, and the head rather large. The dorsal fin is long and usually notched between the spinous anterior fin and the soft-rayed posterior fin; the caudal fin may be rounded, slightly forked, pointed, or somewhere in between, depending on the species. Sciaenids are generally silvery to dark brown (also described as coppery or bronze colored), and may have a dark blotch at the base of the pectoral fin. They are known for making peculiar noises (called croaking, grunting, drumming, or snorting), which apparently involve the swim bladder and are most noticeable during their spawning seasons. They are carnivorous, eating a wide variety of invertebrates and finfish, and generally large: Some reach 10 lb, others over a hundred.

Most are marine fish, usually found on sandy or muddy shores near the mouths of large rivers in warm seas of varying salinity; however, some are endemic to fresh water. Sciaenids live along the eastern seaboard, with some species extending around the northern Gulf of Mexico, or even (for the Atlantic croaker) as far south as Argentina.

REFERENCES

Bigelow, H. G., and W. C. Schroeder. 1953. *Fishes of the Gulf of Maine*. U.S. Fish and Wildlife Service Fishery Bulletin, no. 74. Washington, DC: U.S. Government Printing Office.
Fischer, W. ed. 1978. *FAO species identification sheets for fishery purposes: Western Central Atlantic (fishing area 31)*, vol. 4. Rome: Food and Agriculture Organization of the United Nations.

Aquaculturally Important Fish Species

Bass

SCIENTIFIC AND COMMON NAMES: ***Micropterus salmoides***, **Largemouth bass**
Related: smallmouth bass, *M. dolomieu.*

DESCRIPTION AND DISTINCTIVE CHARACTERISTICS: Olive to brown on the dorsal side, with an irregular, horizontal, faint dark stripe on the sides of the body from the gill cover to the caudal fin. The type of habitat in which largemouth bass live, especially clarity of the water, will cause variation in color.

HABITAT RANGE: Freshwater fish found in all of the southern states of the U.S. mainland; transplanted to Hawaii and several continents. Appear to prefer warm, sluggish waters of lakes and streams.

REPRODUCTION AND DEVELOPMENT: Nest building (making a depression in the substrate) is a responsibility of the male, as is guarding eggs and protecting the numerous young. Females spawn during periods when the water temperature is about 60–65 °F. As many as 40,000 eggs may be laid by a large female.

AGE AND GROWTH: Record largemouth weigh over 25 lb and reach about 32 in. in length.

FOOD AND FEEDING: Eat a wide variety: insects, frogs, crayfish, snakes, mice, fishes (cannibalistic on young bass), and young birds and muskrats.

PARASITES AND DISEASE: Numerous protozoans, trematodes, cestodes, nematodes, acanthocephalans, leeches, mollusks, and crustaceans.

PREDATORS AND COMPETITORS: Smallmouth black bass and other large fishes are competitors and predators. Largemouth bass are cannibalistic.

AQUACULTURAL POTENTIAL: Considered by many devoted freshwater fishers to be the best fish around. The emphasis is on rearing these bass to the fingerling stage for distribution and release into public fishing waters to establish or maintain sportfishing stocks. Large populations of forage fishes can be controlled by the introduction of largemouth bass of about 4 in. in length. One culture method used is called the "spawning–rearing" method, where the adult fish produce fry that remain in the spawning pond until they reach a length of about 1 in. A second method is called the "fry-transfer" method, and is used to produce larger fingerlings in a separate, properly prepared rearing pond. Commercial opportunities exist for hatcheries to sell largemouth bass to state and federal agencies for planting in public waters.

REGIONS WHERE FARMED AND/OR RESEARCHED: Tishomingo National Fish Hatchery, Oklahoma; A. E. Woods State Fish Hatchery, Texas.

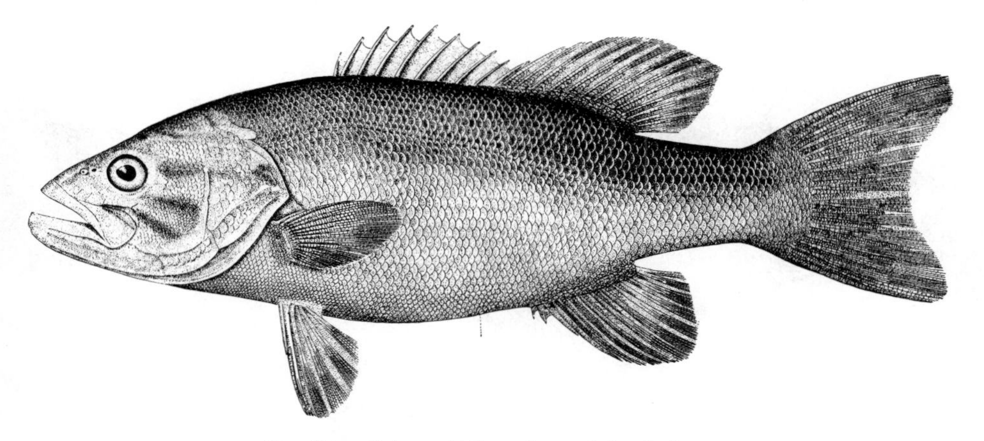

Smallmouth bass (*Micropterus dolomieu*).

REFERENCES

McLarney, W. 1984. *The freshwater aquaculture book; a handbook for small scale fish culture in North America*. Point Roberts, WA: Hartley & Marks.

Regier, H. A. 1963. Ecology and management of largemouth bass and bluegills in farm ponds in New York. *New York Fish and Game Journal* 10(1):1–89.

Stickney, R. R., ed. 1986. *Culture of nonsalmonid freshwater fishes*. Boca Raton, FL: CRC Press.

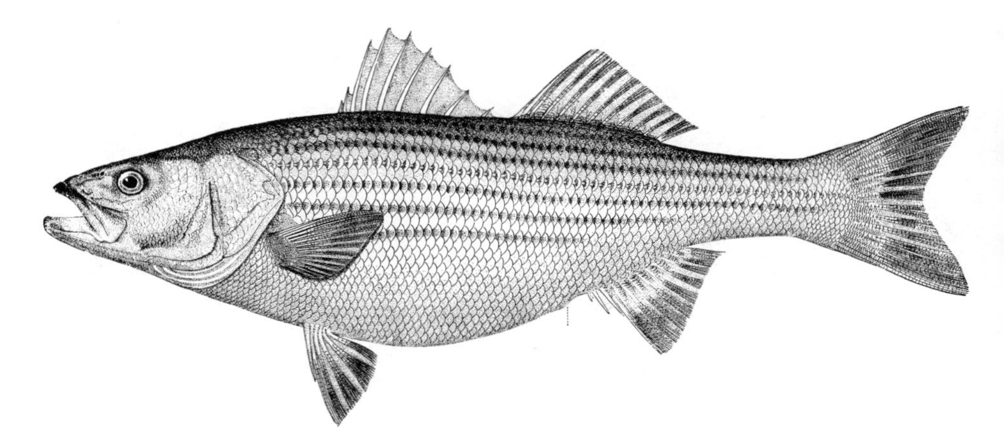

SCIENTIFIC AND COMMON NAMES: ***Morone saxatilis*, Striped bass** Also hybrids × white bass, *M. chrysops* (for which data below vary somewhat).

DESCRIPTION AND DISTINCTIVE CHARACTERISTICS: Striped bass can be identified by horizontal stripes. The head is elongate, and the lower jaw protrudes slightly. The first dorsal fin has spines, the second soft rays; the tail is slightly forked. The back is dark colored, from green to black; silvery sides have seven or eight stripes that extend most of the way along the body.

HABITAT RANGE: Found on both U.S. coasts (common in the San Francisco area) and in the Gulf of Mexico. Lives in fresh, brackish, and salt water; migratory. Landlocked populations exist in inland lakes and reservoirs as results of stocking programs. Temperature range is 56–72 °F.

REPRODUCTION AND DEVELOPMENT: Spawn in fresh or brackish water, late April–early June, laying 65,000 to five million eggs. Moving water is required for spawning because eggs tend to sink. Eggs hatch in 4–72 hr, depending on temperature; about 15 days after hatching, larval bass start feeding.

AGE AND GROWTH: Live about 13 years; at age 3 reach about 15 in., and at age 5 about 20 in., in the Chesapeake Bay area. Recent figures for cultured bass are 0.75–1 lb in 16 months.

FOOD AND FEEDING: Larval striped bass eat zooplankton and/or *Artemia* nauplii in captivity. Adults in natural marine habitats eat fish (eels, herring, flounders, and anchovies) and a variety of crustaceans and other invertebrates. In captivity, fingerlings can easily be trained to eat dry, pelleted food.

DISEASES AND PARASITES: Striped bass in captivity experience diseases common to other fishes held in ponds and cages, such as tail rot and hemorrhagic areas on the fin epidermis. Bacteria and viruses have been reported. Parasites include protozoans, a myxosporean (*Kudoa cerebralis*), a *Trichodina,* and a copepod (*Ergasilus labracis*).

PREDATORS AND COMPETITORS: Predacious insects (e.g., midge larvae), cannibalism, birds, and other fish cause mortality in culture facilities.

AQUACULTURAL POTENTIAL: Aquacultural attempts using striped bass failed in the mid-1970s, but prices are higher now, and improved technology may put this species in a better market position. U.S. striped bass production went from just under 400,000 lb in 1987 to 1.5 million lb in 1991. Hybrid striped bass (especially striped bass × white bass) grow faster, survive better, and resist diseases better than nonhybrids. Striped bass and hybrids generally grow faster in brackish than in fresh water. In early 1990 there were at least forty U.S. hatcheries selling fry and fingerling hybrid striped bass.

REGIONS WHERE FARMED AND/OR RESEARCHED: North and South Carolina, Louisiana, California, Florida, Texas, New York, and Mississippi.

REFERENCES

Bonn, E. W., W. M. Bailey, J. D. Bayless, K. E. Erickson, and R. E. Stevens, eds. 1976. *Guidelines for striped bass culture*. Lawrence, KS: Southern Division, American Fisheries Society.

Hodson, R. G. 1991. Hybrid striped bass culture in ponds. In *CRC Handbook of mariculture, Vol. II: Finfish aquaculture*, ed. J. P. McVey, pp. 167–91. Boca Raton, FL: CRC Press.

Hodson, R. G., and J. Jarvis. 1990. *Raising hybrid striped bass in ponds*. Raleigh, NC: UNC Sea Grant College Program.

McCraren, J. P., ed. 1984. *The aquaculture of striped bass: A proceedings*. Maryland Sea Grant publication no. UM-SG-MAP-84-01. College Park, MD: University of Maryland Sea Grant Program.

Mansueti, R. J., and E. H. Hollis. 1963. *Striped bass in Maryland tidewater*. Solomon, MD: Natural Resources Institute of the University of Maryland.

Porterfield, B. 1981. Innovative New York fish farm cultures, markets striped bass. *Aquaculture Magazine* 8(1):20-22.

Setzler, E. M., W. R. Boynton, K. V. Wood, H. H. Zion, L. Lubbers, N. K. Mountford, P. Frere, L. Tucker, and J. A. Mihursky. 1980. *Synopsis of biological data on striped bass,* Morone saxatilis *(Walbaum)*. NOAA technical report NMFS circular 433. Seattle, WA: National Marine Fisheries Service.

Smith, T. I. J. 1988. Aquaculture of striped bass and its hybrids in North America. *Aquaculture Magazine* 14 (1):40–9.

Smith, T. I. J. 1989. The culture potential of striped bass and its hybrids. *World Aquaculture* 20(1):32–8.

Bowfin

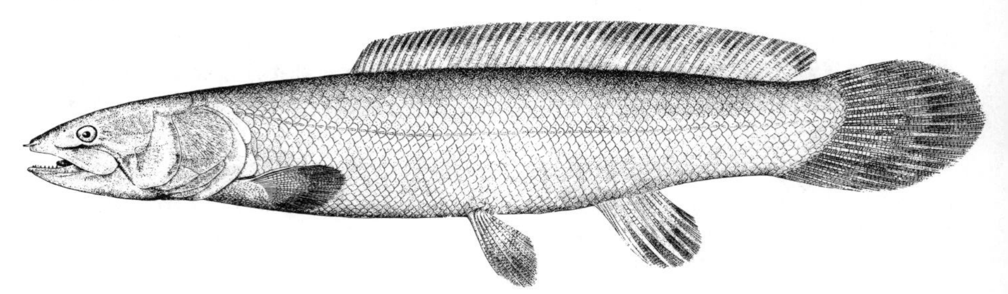

SCIENTIFIC AND COMMON NAMES: ***Amia calva,*** **Bowfin**
Also called dogfish, cotton fish, cypress trout.

DESCRIPTION AND DISTINCTIVE CHARACTERISTICS: Robust, cylindrical body with a blunt bullet-shaped head. A characteristic dark spot with a margin of orange or yellow is present on either side of the tail of males; on females it is not always present or may appear without a margin. A long, wavy dorsal fin extends over almost the entire back.

HABITAT RANGE: U.S. Great Lakes, south to Florida, Mississippi Valley. The bowfin can survive in oxygen-poor habitats by gulping air at the surface of the water.

REPRODUCTION AND DEVELOPMENT: In spring when temperatures reach 60–65 °F, males usually dig a nest in shallow water, guard the eggs, and protect the young. Eggs hatch in about 8–10 days, the young remaining in the nest for about a week.

AGE AND GROWTH: In about six months bowfins may reach roughly 2 lb. Under ideal rearing conditions, bowfins will mature as early as age 1 (4 lb). Maximum weight is about 20 lb.

FOOD AND FEEDING: Opportunistic carnivores, predatory on other fishes, insects, frogs, and oligochaetes. Adults usually eat more fish than juveniles do.

PARASITES AND DISEASE: Numerous species in some groups of the following: trematodes, cestodes, nematodes, Acanthocephala, leeches, crustaceans, and arthropods (Pentastomida).

PREDATORS AND COMPETITORS: Male parents herd their young to protect them from predatory fish.

AQUACULTURAL POTENTIAL: Flesh is not very tasty. One author (Smith 1985) described the fish as having "the delicacy and grace of a stick of firewood." Cooking requires special care to produce a desirable dish. Their large, black eggs are occasionally eaten, and have become known as "Cajun caviar." It is uncertain whether bowfins will eat prepared feeds; if they can be trained to eat pellets in place of live food, rearing will be expedited. Considerably more biological research is needed, as are aquacultural trials. At this time, only a single farm in Indiana offers bowfin for sale.

REGIONS WHERE FARMED AND/OR RESEARCHED: University of Southwestern Louisiana, Lafayette; Louisiana State University, Baton Rouge; Indiana.

REFERENCE

Smith, C. L. 1985. *The inland fishes of New York State*. Albany: N.Y. State Dept. of Environmental Conservation.

Buffalo Fish

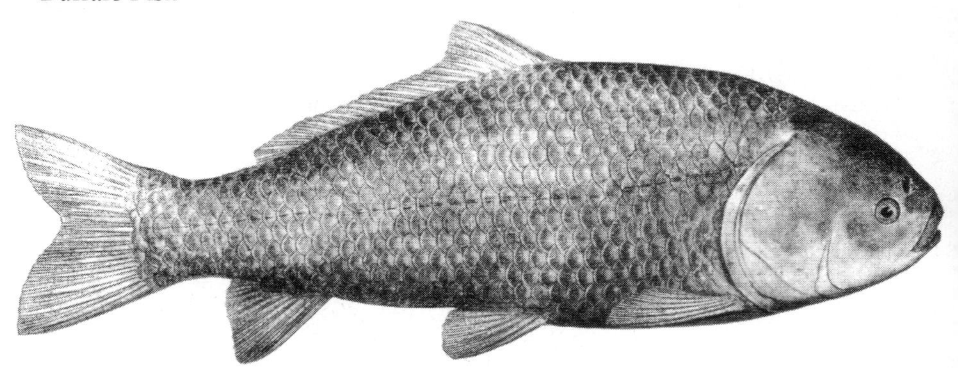

SCIENTIFIC AND COMMON NAMES: ***Ictiobus cyprinellus,* Bigmouth buffalo** Also called by many different common names across its range. Smallmouth buffalo (*I. bubalus*) and black buffalo (*I. niger*) are also used in aquaculture.

DESCRIPTION AND DISTINCTIVE CHARACTERISTICS: Bodies of buffalo fish are robust and elliptical in shape, and usually olive brown or coppery in color. The head is large, the snout blunt and broadly rounded, the mouth large, wide, and protractile forward. Buffalo fish may grow quite large: One specimen of smallmouth found in the upper Mississippi was over 4 ft long and weighed 65 lb.

HABITAT RANGE: In slow-moving U.S. waters from North Dakota to Lake Erie and southward to Alabama and Texas. Common in the Mississippi and Minnesota rivers.

REPRODUCTION AND DEVELOPMENT: Buffalo fish spawn from late in April into June. A 14-lb female may produce 750,000 adhesive eggs, which attach to plants and debris. They hatch after about 10 days at a temperature of 62 °F.

AGE AND GROWTH: May reach 20 years of age and weigh as much as 80 lb in southern states; usually marketed at 2–7 lb.

FOOD AND FEEDING: Eat mollusks, amphipods, vegetation, and aquatic insect larvae. Zooplankton is important in the diet of the bigmouth buffalo. In polyculture, they will also feed on excess feed supplied to other species.

PARASITES AND DISEASE: Trematodes, cestodes, nematodes, Acanthocephala, leeches, and crustaceans.

PREDATORS AND COMPETITORS: Bigmouth buffalos compete with carp, suckers, and smallmouth buffalo for food.

AQUACULTURAL POTENTIAL: Of the three species of buffalo fish native to North America, the bigmouth buffalo is the best candidate for farming. Ease of adaptation to lake and pond environments, rapid growth, early maturity, and high fecundity are attributes that give this species the edge. Research has tested polyculture combinations of bigmouth buffalo with channel catfish, crayfish, paddlefish, tilapia, and hybrids of bigmouth × smallmouth buffalo. Buffalo fish may also be raised in monoculture in rice fields, and can be stripped of their sex products so that the eggs can be fertilized in the hatchery. On the negative side, they resemble carp, and their flesh has many small bones. Hybridization experiments using bigmouth and black crosses produced fish twice the length and four times the weight of nonhybrids.

REGIONS WHERE FARMED AND/OR RESEARCHED: Arkansas.

REFERENCES

Becker, G. C. 1983. *Fishes of Wisconsin.* Madison: University of Wisconsin Press.

McLarney, W. 1984. *The freshwater aquaculture book; a handbook for small scale fish culture in North America.* Point Roberts, WA: Hartley & Marks.

Stickney, R. R., ed. 1986. *Culture of nonsalmonid freshwater fishes.* Boca Raton, FL: CRC Press.

Carp

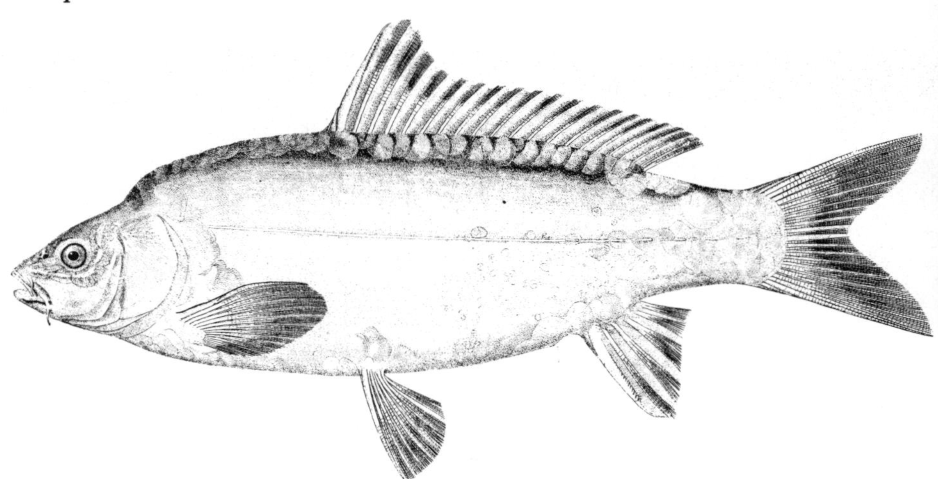

SCIENTIFIC AND COMMON NAMES: ***Cyprinus carpio,*** **Carp**
Also called common carp. Based on the prevalence or absence of scales, several varieties (some may be subspecies) are recognized: scale carp, mirror carp, and leather carp. Body shape may also distinguish some varieties. The koi is an ornamental strain of carp that exhibits a wide range of colors arranged in patterns (see Chapter 4 under "Ornamental species").

DESCRIPTION AND DISTINCTIVE CHARACTERISTICS: A large, robust fish whose mouth has two pairs of *barbels* (slender tactile structures, or "feelers"). A single stout spine precedes the long dorsal fin. The amount and arrangement of scales varies, as does coloration.

HABITAT RANGE: Freshwater fish of Asia; transplanted widely first into Europe then into the United States.

REPRODUCTION AND DEVELOPMENT: The male carp matures at about age 2 and the female at 3. In temperate regions the fish spawn once a year, in spring, usually when water temperatures reach 64–68 °F. Carp will spawn five or six times a year in tropical waters. The adhesive eggs are fertilized as they are released and attach to vegetation; hatching takes place within two to six days, and the larvae cling to aquatic plants for several more days.

AGE AND GROWTH: Wild carp may live about 20 years, but ages of 70–80 have been recorded. Weights of 35–40 lb have been recorded frequently, but the average is closer to 10–15 lb. Carp reach over 2 ft in length.

FOOD AND FEEDING: Carp are classified as omnivorous. They eat aquatic insects, rotifers, and copepods on plants in ponds and lakes when young, rooting in the bottom and destroying vegetation as they get older. Their dentition permits them to crack shell and feed on some of the shelled invertebrates. They require artificial feeds for maximum production in culture facilities.

PARASITES AND DISEASE: Bacterial diseases (infectious carp dropsy, infectious abdominal dropsy, columnaris disease); "Ich" or white spot disease, a problem in culture, due to the protozoan *Ichthyophthirius;* and external monogenic trematodes (*Dactylogyrus* and *Gyrodactylus*).

PREDATORS AND COMPETITORS: Carp compete for food with some native freshwater fishes and may destroy the habitat used by other fishes.

AQUACULTURAL POTENTIAL: There is a considerable body of literature on the biology and culture of carp. They can be raised in net pens, in ponds with little circulation, and in flowing water. Wild carp in the New World suffer from an image of a poor man's or "trash" fish; however, when properly prepared it is an excellent dish and has enjoyed a fine market in Asia and Europe for centuries. The common carp supports a large commercial industry in Japan, where early carp culture was combined with rice growing (rice-paddy culture). This involved home-made feeds, whose manufacture was a laborious task; today, artificial feeds are available. The Chinese have been very successful at integrated polycultural fish farming, raising several species of fish (including different species of carp) together with land plants, ducks, and other products.

REGIONS WHERE FARMED AND/OR RESEARCHED: Japan, Indonesia, China, Europe, Israel, and the United States.

REFERENCES

Cooper, E. L., ed. 1987. *Carp in North America.* Bethesda, MD: American Fisheries Society.
Michaels, V. K. 1988. *Carp farming.* Farnham (U.K): Fishing News Books.

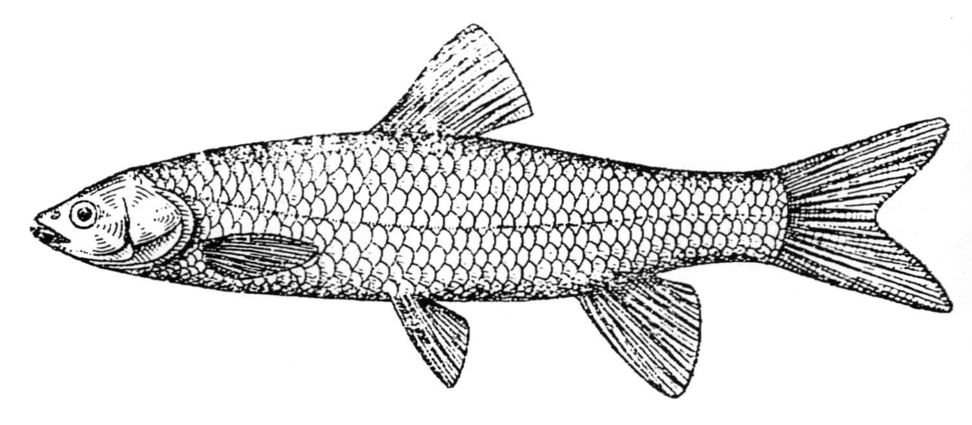

SCIENTIFIC AND COMMON NAMES: ***Ctenopharyngodon idella,* Grass carp** Also called white amur.

DESCRIPTION AND DISTINCTIVE CHARACTERISTICS: A large, rather typical carp, with pharyngeal teeth for eating aquatic vegetation, the grass carp is a fast swimmer and strong jumper. Its head is broad, the mouth subterminal, and the belly rounded. The body is covered with large, heavy scales and varies in color from light silver to bronze, being somewhat darker on its dorsal side.

HABITAT RANGE: A native of Chinese lowland rivers and the Amur River (Russian–Mongolian border). Imported from Asia to many countries, including the United States; introduced for culture, experimentation, and weed control.

REPRODUCTION AND DEVELOPMENT: Males mature at about 1 year old, females at about age 2. Normally spawn during the monsoon season, with fecundity as high as two million eggs per female. Grass carp will not spawn in ponds with static water. Because of the concern of conservation organizations that transplanted grass carp would reproduce rapidly and compete seriously with local species for their food and space, sterile *triploid* fishes (i.e., those with three sets of chromosomes) have been grown from recently fertilized grass carp eggs treated with the chemical cytochalasin or exposed to heat or cold shock.

AGE AND GROWTH: Growth is rapid, with wild fish reaching a length of 6–12 in. by age 1. In culture and in different areas, growth varies greatly and is a function of density, feed, competition with other fish, and the suitability of environmental conditions. Grass carp may attain a weight of 100 lb.

FOOD AND FEEDING: Small grass carp eat rotifers and protozoan phytoplankton; at lengths above about 0.1 in. their natural diet is macrovegetation. They are known to consume 100 percent of their body weight daily and to digest only about 50 percent of the plants ingested. Their diet as adults has encouraged their transplantation to areas where natural waterways and lakes have become choked by weeds, and for culture as food fish. When vegetation suitable for food is limited, they become omnivorous.

PARASITES AND DISEASE: Viruses, bacteria, fungi, protozoans, trematodes, cestodes, nematodes, crustaceans, and pentastomids.

PREDATORS AND COMPETITORS: Compete with carp and game fish (bluegills and bass), and adversely alter the ecology of an area by removing vegetation used for protection by the young of some fish.

AQUACULTURAL POTENTIAL: Grass carp were imported for biological control of noxious aquatic vegetation. Tests of their organoleptic and sportfishing qualities suggest that they have limited potential as a game (fighting) fish but are acceptable for table use. Juveniles and adults can be fed kitchen refuse, cereal brans, oilcakes, and silkworm pupae; night soil and dung are used in certain countries. To avoid competition with indigenous species, sterile (triploid) grass carp are used for weed control. Laws prohibit hatching of grass carp in Arizona, since the escape of nonsterile grass carp from a hatchery could damage the local ecology.

REGIONS WHERE FARMED AND/OR RESEARCHED: U.S. Southeast (Arkansas, Missouri, Alabama), Nebraska, and California.

REFERENCES

Jhingran, V. G., and R. S. V. Pullin. 1985. *A hatchery manual for the common, Chinese and Indian major carps.* Manila: Asian Development Bank and ICLARM.

Shireman, J. V., and C. R. Smith. 1983. *Synopsis of biological data on the grass carp,* Ctenopharyngodon idella *(Cuvier and Valenciennes, 1844).* FAO Fisheries Synopsis no. 135. Rome: Food and Agriculture Organization of the United Nations.

Zonneveld, N., and H. V. Zon. 1985. The biology and culture of grass carp (*Ctenopharyngodon idella*), with special reference to their utilization for weed control. In *Recent advances in aquaculture,* vol. 2, ed. J. F. Muir and R. J. Roberts, pp. 119–91. London: Croom Helm.

Catfish

SCIENTIFIC AND COMMON NAMES: ***Ictalurus punctatus*, Channel catfish**

DESCRIPTION AND DISTINCTIVE CHARACTERISTICS: Distinguishable from many other species of finfish by long barbels ("feelers") around the mouth, which are used for locating food. Catfish have heavy, sharp pectoral and dorsal spines. The body is scaleless with a deeply forked tail.

HABITAT RANGE: Warm (70–84 °F) fresh waters in the southeastern United States. Catfish appear to prefer moving water in channels, but will survive in still water.

REPRODUCTION AND DEVELOPMENT: Channel catfish will spawn in ponds, laying 15,000–38,000 eggs in a nest built in a receptacle. Male catfish care for eggs after spawning; incubation lasts 5–10 days. After hatching the young stay on the bottom for about two days and, usually by the third day, will begin feeding and moving about.

AGE AND GROWTH: Catfish usually weigh about 1 lb after 14–18 months. A record channel catfish weighed 57 lb.

FOOD AND FEEDING: Omnivorous, a scavenger and predator. In culture facilities, commercially available pelleted feeds are used. Fish meal or ground meat scraps (tankage) seems to be essential in all catfish feeds.

PARASITES AND DISEASES: Viruses, bacteria, a protozoan ("Ich" or white spot disease), monogenetic trematodes, cestodes (tapeworms); also brown blood disease. Many noninfectious diseases kill cultured catfish. As many as 62 different species of macroparasites are known to parasitize channel catfish.

PREDATORS AND COMPETITORS: Predacious insects, frogs, snakes, and larger catfish eat the fry. Many birds, including cormorants, herons, egrets, and kingfishers, can cause serious losses in catfish ponds.

AQUACULTURAL POTENTIAL: Although a very popular food fish in the South, channel catfish was considered a poor man's fish in other parts of the United

States. That image was successfully eroded by the introduction of catfish to fast-food restaurants. Production of farmed catfish has increased dramatically in recent years to a point where the U.S. market may be near saturation; to maintain falling farm gate prices, markets outside of the United States may have to be developed. Catfish are often raised as an additional crop with agriculture crops such as cotton, rice, and soya. Occasionally, catfish may be unsuitable for market due to an off-flavor (muddy or musty taste) caused by blue-green algae in ponds. Artificial selection could produce better food conversions, more rapid growth, and greater tolerance to low dissolved oxygen. In some types of farm ponds and/or at high fish densities, aeration is required to keep the fish in good condition.

REGIONS WHERE FARMED AND/OR RESEARCHED: Southeastern United States.

REFERENCES

Johnson, J. M. 1971. *Catfish farming handbook*. San Angelo, TX: Educator Books.

Lee, J. S. 1981. *Commercial catfish farming*, 2nd ed. Danville, IL: Interstate.

Louisiana Agricultural Experiment Station, Louisiana Cooperative Extension Service. 1988. *Commercial production of farm-raised catfish*. Baton Rouge: Louisiana State University Agricultural Center.

Tucker, C. S., ed. 1985. *Channel catfish culture*. Developments in Aquaculture and Fisheries Science, vol. 15. Amsterdam: Elsevier.

COMMON AND SCIENTIFIC NAMES: ***Clarias batrachus,* Walking catfish**

DESCRIPTION AND DISTINCTIVE CHARACTERISTICS: This fish resembles an eel and is able to breath air. Normal coloration is gray, brown, or black. Males have a dark blotch on the dorsal fin and are more brightly colored than females. Many are albinos with a dark pink head and pale pink over the rest of the body. As with other catfish, barbels (8) are present on a large flattened head. They are able to move over land by wiggling their bodies, aided by pectoral fins with stout spines.

HABITAT RANGE: Naturally distributed in eastern India and Southeast Asia and the Philippines; transplanted to southern Florida and many tropical and subtropical countries. These catfish live in fresh and brackish waters (68 °F) and can survive in temperatures as high as 90 °F.

REPRODUCTION AND DEVELOPMENT: Eggs are deposited in the nest by the females and guarded by males.

AGE AND GROWTH: May reach 22 in. in length (one report says 16 in.). Can be grown to market size of 5.25–10 lb in four to six months.

FOOD AND FEEDING: Eat water plants, worms, mussels, fishes, and even insects. Farmed walking catfish are usually fed trash fish.

PARASITES AND DISEASES: The bacterium *Aeromonas hydrophila,* a protozoan (*Trichodina* spp.), and a monogenetic trematode (*Gyrodactylus* spp.).

PREDATORS AND COMPETITORS: Wading birds and larger fish prey on walking catfish.

AQUACULTURAL POTENTIAL: Walking catfish are very hardy and can live in muddy, low-oxygen water (they possess an accessory breathing apparatus). They can be cultured at high densities, resulting in high production per unit water area. Ponds must be enclosed, since this species can move short distances over land: Walking catfish have received considerable notoriety escaping from tropical fish dealers and becoming established in Florida, where it is believed they can compete successfully against local, less hardy species. Despite their rearing advantages, the aquacultural potential of walking catfish is somewhat limited due to legal problems associated with introducing an exotic species, the need for year-round warm water, the comparative toughness of the meat, and the lack of an established U.S. market.

REGIONS WHERE FARMED AND/OR RESEARCHED: Thailand, China, Africa.

REFERENCES

Areerat, S. 1987. *Clarias* culture in Thailand. *Aquaculture* 63:355–62.

Burgess, W. 1989. *An atlas of freshwater and marine catfishes; a preliminary survey of the Siluriformes*. Neptune City, NJ: T.F.H. Publications.

Diana, J. S., S. L. Kohler, and D. R. Ottey. 1988. A yield model for walking catfish production in aquaculture systems. *Aquaculture* 71:23–35.

Idyll, C. P. 1969. New Florida resident, the walking catfish. *National Geographic* 135:846–51.

Char

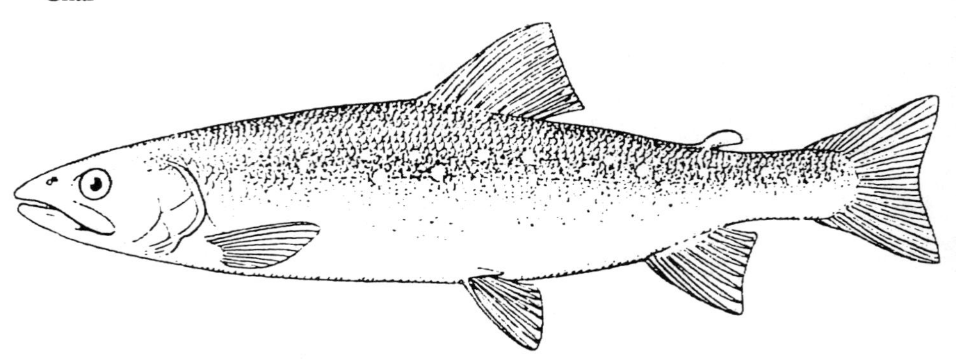

SCIENTIFIC AND COMMON NAMES: ***Salvelinus alpinus,* Arctic char** There are about a dozen or so other common names, including Arctic salmon, sea trout, Hudson Bay salmon, and Greenland char.

DESCRIPTION AND DISTINCTIVE CHARACTERISTICS: Similar to other trout species. Sea-run fish are bright silver, but freshwater spawners are bright greenish, bluish, or orange-red. The systematics of Arctic char are in a state of confusion due to the wide variation in characters, ecology, and growth rates from different regions.

HABITAT RANGE: Anadromous, spending juvenile period (5–7 yr) in rivers or lakes, then moving to sea, but returning to fresh water to spawn. In some areas, Arctic char are landlocked, and thus spend their entire lives in fresh water. Distribution is circumpolar in nearshore marine waters. They also occur in northern Asia, Northern Europe, Greenland, and Iceland. Landlocked char may be found farther south than their anadromous counterparts.

REPRODUCTION AND DEVELOPMENT: Spawning occurs during the fall, usually September–December, depending on location. Females build nests in gravel bottoms of lakes or streams when temperatures are about 39 °F, and lay 3,000–5,000 eggs. Females in northern populations spawn only every second or third year.

AGE AND GROWTH: Char grow rather slowly, though those that go to sea tend to grow faster and reach a larger size than those that are landlocked. They may reach 40 years of age. A 20-year-old fish might be about 28 in. long. Char will reach pan size (8–10 oz.) in 12–16 months.

FOOD AND FEEDING: Arctic char are carnivores and eat a wide variety of marine and freshwater fish and invertebrates. In one study, 34 species of food items were found in the stomachs of char.

PARASITES AND DISEASE: Over 37 species of parasites have been described from the Arctic char, representing a variety of different groups from protozoans to worms.

PREDATORS AND COMPETITORS: Loons and terns eat small char, and seals and fish prey on large sea-run char.

AQUACULTURAL POTENTIAL: Fine flavor is a plus, but slow growth, requiring large amount of rather expensive feeds, reduces the profit margin. The Arctic char is a rather efficient converter because it feeds both in the water column and on the bottom of holding tanks (many salmonid fishes feed only in the water column). Experiments using Arctic char and rainbow trout for comparative studies of growth and stocking densities show that the char is better suited to culture at high densities, provided water quality is high. Concern has been expressed over transplanting the Arctic char outside of its normal range.

Most biological research and farming trials for this cold-water species have been conducted outside the United States. In Canada, there has been some concern over increased Arctic char production, which may adversely affect the traditional char capture fisheries' supply and demand.

REGIONS WHERE FARMED AND/OR RESEARCHED: Anchorage, Alaska; Canada; Northern Europe, Norway, Sweden, and Iceland.

REFERENCES

Balon, E. K., ed. 1980. *Charrs, salmonid fishes of the genus* Salvelinus. The Hague: W. Junk.

Holm, J. C. 1989. Mono- and duoculture of juvenile Atlantic salmon (*Salmo salar*) and Arctic char (*Salvelinus alpinus*). *Canadian Journal of Fisheries and Aquatic Sciences* 46(4): 697–704.

Johnson, L., and B. Burns, eds. 1984. *Biology of the Arctic charr: Proceedings of the International Symposium on Arctic Charr*. Winnipeg: University of Manitoba Press.

Scott, W. B., and M. G. Scott. 1988. *Atlantic fishes of Canada*. Canadian Bulletin of Fisheries and Aquatic Sciences no. 219. Toronto: University of Toronto Press.

Cod

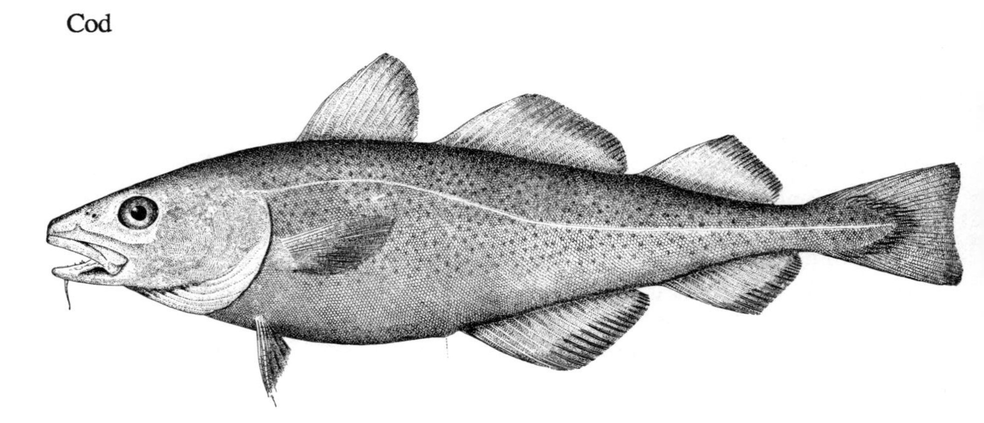

SCIENTIFIC AND COMMON NAMES: ***Gadus morhua,* Atlantic cod**

DESCRIPTION AND DISTINCTIVE CHARACTERISTICS: Similar to hake and haddock, with three separate dorsal fins and two anal fins. It is further characterized by a large head, stocky body, blunt nose, and wide mouth. Although similar to haddock, cod can be distinguished by a pale lateral line. They differ from pollack in having a squarish caudal fin, projecting upper jaw, and mottled color pattern.

HABITAT RANGE: In American waters, from the southern tip of Baffin Island to North Carolina. Also in coastal waters of Greenland, and along the coast of Europe. Live in cold waters, in close association with the bottom, over a wide range of depths (1–250 fathoms).

REPRODUCTION AND DEVELOPMENT: A winter spawner with peaks in December. Females reach maturity at age 2–12 years, depending on geographic location of the stocks; they are very fecund, laying up to 9 million eggs per female.

AGE AND GROWTH: May reach age 30. There is a single record of a cod 6 ft long and weighing 200 lb, and several of cod weighing over 100 lb.

FOOD AND FEEDING: Cod are bottom feeders. Adults eat shellfish, clams, mussels, crabs, squid, and small fish.

PARASITES AND DISEASE: Bacteria, coccidians (protozoans), nematodes (cod worms), and cestodes. Noninfectious diseases or those of unknown cause include eye disease and tumors.

PREDATORS AND COMPETITORS: Seals and fishes are important predators.

AQUACULTURAL POTENTIAL: Cod is Norway's main roundfish, and the fishery is of great economic value. Cod fry can be reared artificially with reasonable survival; however, rearing young stages is rather expensive. Atlantic cod take to artificial diets and are good converters. They grow quite well in the low sea temperature in and around the Norwegian and Scottish waters. If capture fishery stocks decline further and the price increases, the stage may be set for extensive farming of this species. If not, the outlook for commercially successful aquacultural ventures at this time is rather bleak.

REGIONS WHERE FARMED AND/OR RESEARCHED: Norway and Newfoundland. Early trials in Newfoundland and the United States.

REFERENCES

Dahl, E., D. S. Danielssen, E. Moksness, and P. Solemdal. 1984. *The propagation of cod,* Gadus morhua, *L.: An international symposium. Arendal 14-17 June l973.* Arendal, Norway: Institute of Marine Research, Flodevigen Biological Station.

Huse, I. 1991. Culturing of cod (*Gadus morhua* L.). In *CRC handbook of mariculture, Vol. II: Finfish aquaculture,* ed. J. P. McVey, pp. 43–50. Boca Raton, FL: CRC Press.

Jones, A. 1984. Does cod farming have a commercial future? A technical and economic assessment. In *The propagation of cod* Gadus morhua *L: An international symposium, Arendal, 14-17 June 1983,* ed. E. Dahl, D. S. Danielssen, E. Moksness, and P. Solemdal, pp. 773–85. Arendal, Norway: Institute of Marine Research, Flodevigen Biological Station.

Knutsen, G. M., and S. Tilseth. 1985. Growth, development, and feeding success of Atlantic cod larvae *Gadus morhua* related to egg size. *Transactions of the American Fisheries Society* 114:507–11.

Oiestad, V., P. G. Kvenseth, and A. Folkvord. 1985. Mass production of Atlantic cod juveniles *Gadus morhua* in a Norwegian saltwater pond. *Transactions of the American Fisheries Society* 114:590–5.

Crappie

SCIENTIFIC AND COMMON NAMES: ***Pomoxis annularis,* White crappie**
Related: black crappie, *P. nigromaculatus.*

DESCRIPTION AND DISTINCTIVE CHARACTERISTICS: The crappies are centrarchids (members of the sunfish family). They have rather deep bodies with two dorsal fins joined together; the anterior fin has sharp spines. The white crappie is silver colored with dark bars on the sides of the body.

HABITAT RANGE: Widely distributed in U.S. freshwater lakes and streams. It is found as far north as southern Minnesota and as far south as the Gulf of Mexico, and also inhabits Alabama and South Carolina.

REPRODUCTION AND DEVELOPMENT: May spawn at 1–2 years old. As many as 140,000 eggs may be released from large (1.5-lb) females. Spawning occurs during late spring–early summer, when water temperatures reach 65 °F. Crappie nests may be built on soft, muddy bottoms.

AGE AND GROWTH: Infrequently large adults weighing up to about 4 lb have been reported; however, most usually weigh 1–2 lb. In culture, few white crappies live longer than about four years.

FOOD AND FEEDING: Eat aquatic insects, crustaceans, and small fish (e.g., minnows). When *Daphnia* are very abundant, white crappie may eat large numbers of them. Small white crappie will begin taking pelleted food after a short training period. In aquaculture, some cannibalism by adults on young has been observed.

PARASITES AND DISEASE: Bacterial disease (columnaris = saddle back disease) is rather common in culture. In wild specimens: protozoans, trematodes, cestodes, nematodes, acanthocephalans, leeches, and crustaceans.

PREDATORS AND COMPETITORS: Crayfish are known to eat crappie eggs.

AQUACULTURAL POTENTIAL: There are optimistic projections for the farming of some of the sunfish family species because of their good taste and popularity as sportfish. Important aquacultural aspects still need to be studied, the most important being economics of their culture, feeds, and improvement of stocks through artificial selection. Black crappie are considered to be superior to white crappie for rearing in small ponds and for their tolerance to handling.

REGIONS WHERE FARMED AND/OR RESEARCHED: Centerton State Fish Hatchery, Arkansas; Colorado State University, Fort Collins; Texas Parks and Wildlife, Austin; Illinois Natural History Survey.

REFERENCES

McLarney, W. 1984. *The freshwater aquaculture book; a handbook for small scale fish culture in North America.* Point Roberts, WA: Hartley & Marks.

Martin, M. 1988. Black and hybrid crappie culture and crappie management. *Aquaculture Magazine* 14(3):35-41.

Stickney, R. R., ed. 1986. *Culture of nonsalmonid freshwater fishes.* Boca Raton, FL: CRC Press.

Croaker

SCIENTIFIC AND COMMON NAMES: ***Micropogonias undulatus*, Atlantic croaker**

DESCRIPTION AND DISTINCTIVE CHARACTERISTICS: This medium-sized fish is silvery with a pinkish cast; back and upper sides are grayish with numerous black spots. The first dorsal fin spiny, the second long with soft rays. Mouth is inferior.

HABITAT RANGE: Euryhaline, inshore waters out to about 100 ft from Massachusetts around the Gulf of Mexico south to Argentina. Early life is spent in and around estuaries.

REPRODUCTION AND DEVELOPMENT: Spawn near and in estuaries during fall. After about age 1 year, croakers leave estuaries.

AGE AND GROWTH: 6.3 in. at age 1 year, 8.7 in. at age 2, 10.9 in. at age 3; apparently, very few croakers live longer than this.

FOOD AND FEEDING: Carnivorous, benthic feeders. Stomachs of young croakers contain about 50 percent copepods; also eat annelid worms.

PARASITES AND DISEASE: Virus (lymphocystis); bacteria may cause mortalities in cage culture; gas bubble disease.

PREDATORS AND COMPETITORS: Predatory fishes.

AQUACULTURAL POTENTIAL: Croakers are high in protein and low in fat; however, they are costly to feed, and eat high on the food web. Croakers have not been spawned artificially, but may be successful in polyculture facilities. Rearing trials in heated power-plant effluent have caused gas bubble disease unless cages are anchored below the surface. Market demand is generally low: Its value as a species for commercial aquaculture must await additional biological data and rearing trials.

REGIONS WHERE FARMED AND/OR RESEARCHED: Louisiana State University, Baton Rouge; Auburn University, Auburn, Alabama; Gulf Coast Research Laboratory, Ocean Springs, Mississippi; Texas A&M University, College Station.

REFERENCES

Austin, C. B., J. C. Davis, R. D. Brugger, and J. A. Browder. 1978. *Croaker workshop report and socio-economic profile*. University of Miami Sea Grant Special Report no. 16. Coral Gables: Sea Grant, University of Miami.

Avault, J. W., C. L. Birdsong, and W. G. Perry. 1970. Growth, survival, food habits, and sexual development of croaker, *Micropogon undulatus*, in brackish water ponds. *Proceedings of the 23rd Annual Conference, Southeastern Association of Game and Fish Commissioners, 1969*, pp. 251–5.

Dolphin Fish

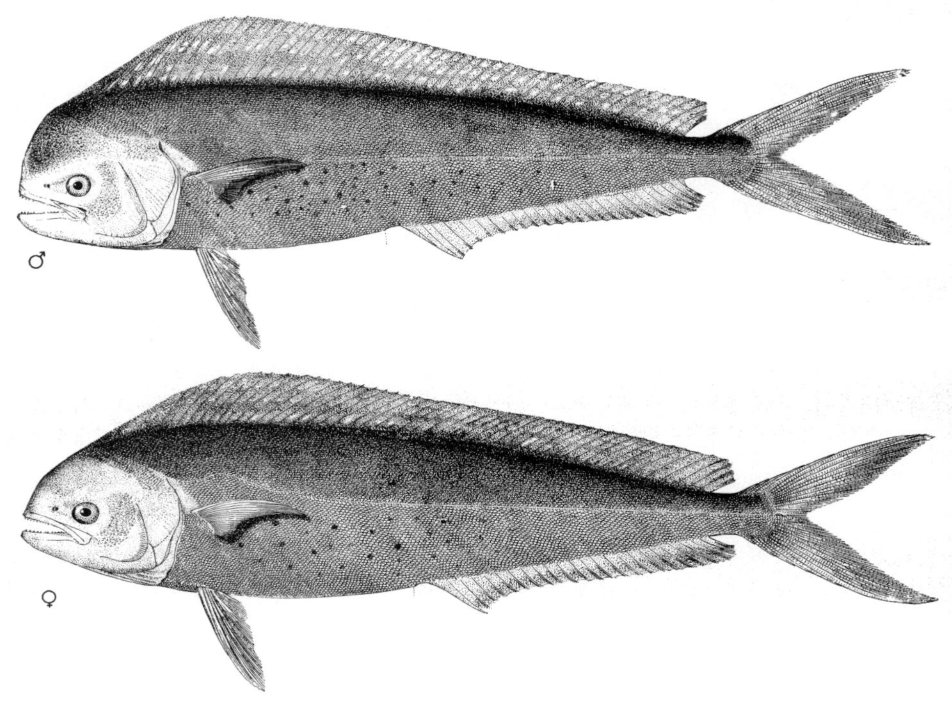

SCIENTIFIC AND COMMON NAMES: ***Coryphaena hippurus*, Common dolphin**
Also called mahimahi.
Related: Pompano dolphin, *C. equisetis*.

DESCRIPTION AND DISTINCTIVE CHARACTERISTICS: A very colorful fish, iridescent blue or blue green on the back, bluish gold on the sides, and yellow or whitish on the belly. Spots on the sides are blue, black, or gold. The body is elongate, laterally compressed, and has long dorsal and anal fins. The forehead is blunt, especially in large males, though more sloped in females (see illustration above). The tail is deeply forked.

HABITAT RANGE: Distributed worldwide in tropical and subtropical waters. Common in the Gulf of Mexico and in the warm waters of the Gulf Stream and Hawaii. Dolphin fish have been seen as far north as Nova Scotia and as far south as South Africa.

REPRODUCTION AND DEVELOPMENT: Female dolphin fish produce 80,000 to one million eggs per spawning, and spawn two or three times a year. Dolphin fish have mated and spawned in captivity; the eggs were collected and reared to juvenile and adult stages.

AGE AND GROWTH: Dolphin fish grow very rapidly: There are records of going from egg to over 20 lb in one year. In the Miami Seaquarium, dolphin grew from 1 lb to 36 lb in eight months. They probably only live about four years.

FOOD AND FEEDING: Juvenile sailfish, flying fish, squid, shrimp, and land crabs are eaten by dolphin fish in nature. In captivity, they eat pelleted food and various kinds of cut-up fish. A conversion ratio of about 3.5 lb of food to 1 lb of dolphin has been achieved.

PARASITES AND DISEASE: Nematodes, isopods, acanthocephalans, trematodes, and copepods (ectoparasites) parasitize dolphin fish.

PREDATORS AND COMPETITORS: Eaten by tunas, marlin, swordfish, and sailfish. Larval swordfish eat dolphin larvae.

AQUACULTURAL POTENTIAL: Their rapid growth, spawning in captivity, fine taste, and rather wide tolerance to environmental conditions make these fish desirable for farming. However, since they are top-level predators, with a low food intake–fish flesh conversion rate, they are expensive to feed. Marketability may be limited in some areas. In colder climates rearing may be restricted to about six months during the warmest part of the year.

REGIONS WHERE FARMED AND/OR RESEARCHED: Oceanic Institute, Honolulu, Hawaii; Harbor Branch Foundation, Fort Pierce, and University of Miami, Florida; North Carolina State University, Raleigh; Southern Sea Farms, Ltd., Perth, Australia.

REFERENCES

Hassler, W. W., and W. T. Hogarth. 1977. The growth and culture of dolphin, *Coryphaena hippurus*, in North Carolina. *Aquaculture* 12:115–22.

Kraul, S. 1991. Hatchery methods for the mahimahi, *Coryphaena hippurus*, at Waikiki Aquarium. In *CRC handbook of mariculture, Vol. II: Finfish aquaculture*, ed. J. P. McVey, pp. 241–50. Boca Raton, FL: CRC Press.

Palko, B. J., G. L. Beardsley, and W. J. Richards. 1982. *Synopsis of the biological data on dolphin-fishes,* Coryphaena hippurus *and* Coryphaena equiselis *Linnaeus*. NOAA Technical Report NMFS CIRC 443. Seattle: National Marine Fisheries Service.

Schekter, R. C. 1983. Mariculture of dolphin (*Coryphaena hippurus*): Is it feasible? *Proceedings of the Gulf and Caribbean Fisheries Institute* 35:27–32.

Szyper, J. P. 1991. Culture of mahimahi: Review of life stages. In *CRC handbook of mariculture, Vol. II: Finfish aquaculture*, ed. J. P. McVey, pp. 228–40. Boca Raton, FL: CRC Press.

Drum

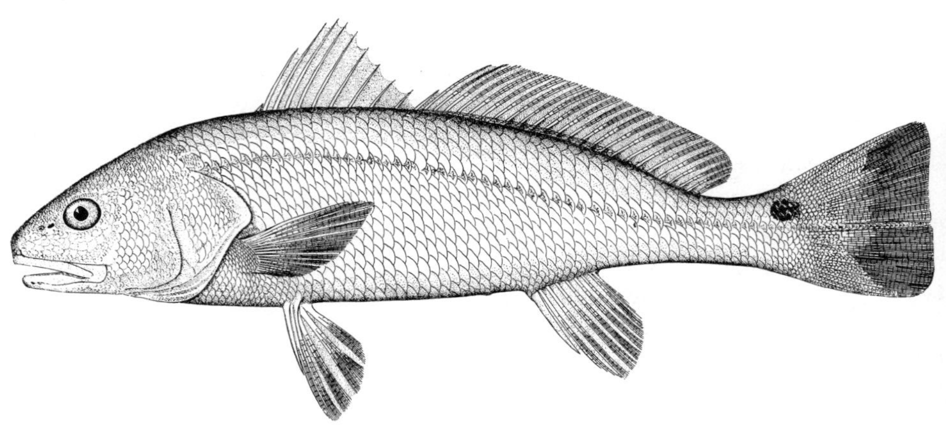

SCIENTIFIC AND COMMON NAMES: ***Sciaenops ocellatus,* Red drum**
Also called redfish, channel bass. Related: *Pogonias cromis,* black drum.

DESCRIPTION AND DISTINCTIVE CHARACTERISTICS: Coppery or bronze colored, usually with a rather prominent black spot on the tail. Upper jaw projects beyond the lower. It may reach a length of 150 in. and weigh 83 lb.

HABITAT RANGE: Mexico north to Cape Cod; usually found in oyster bars, mud and grass flats, and channels. Tends to be nonmigratory. Red drums are sensitive to rapid drops in temperature; their normal range is about 50–96 °F.

REPRODUCTION AND DEVELOPMENT: A large female will release about half a million eggs per spawning.Techniques to spawn and rear red drums artificially have existed since about 1975. More than one spawning annually has been achieved by controlling the photoperiod. Age at first maturity is about 4 years.

AGE AND GROWTH: Growth rates vary considerably. In Texas, red drum fed a high density of forage fish (tilapia) reached 2–3 lb in the first year, 8 lb in the second, 14–15 lb in the third, and 20 lb in the fourth.

FOOD AND FEEDING: Diet includes polychaetes, stomatopods, amphipods, decapods, mollusks, echinoderms, and fishes—mainly fish and crabs.

PARASITES AND DISEASE: Wild fish may be rather heavily parasitized by a wide range of organisms from protozoans to crustaceans. Over 30 species of parasites have been reported; copepods and cestodes occur most frequently.

PREDATORS AND COMPETITORS: Large fish and wading birds will eat small red drum.

AQUACULTURAL POTENTIAL: The potential for red drum seems greatest in warm brackish or saltwater ponds. It has been extensively researched (primarily in Texas) for stock enhancement and, more recently, private aquaculture. These efforts have shown that red drum are tolerant to extremes in environmental conditions, lack obvious disease problems in enclosures, and are quick to accept pelleted food. They can probably be raised to a market size in about a year. A strong market demand developed with the introduction of "blackened redfish," a Cajun dish. Crossing red drum with black drum has improved the growth of the latter species, which also appears to have some aquacultural potential.

REGIONS WHERE FARMED AND/OR RESEARCHED: Gulf Coast Research Laboratory, Ocean Springs, Mississippi; Louisiana State University Agricultural Experiment Station, Baton Rouge; Lyle S. St. Amant Marine Laboratory, Louisiana; Texas A&M University, College Station; Texas Parks and Wildlife Department, Coastal Fisheries Branch; John Wilson Marine Fish Hatchery, Corpus Christi, Texas; University of Miami, RSMAS, Florida; Martinique, French West Indies.

REFERENCES

Arnold, C. R., G. J. Holt, and P. Thomas, eds. 1988. *Red drum aquaculture; proceedings of a symposium on the culture of red drum and other warm water fishes, Corpus Christi, TX, 1987*. Port Aransas, TX: University of Texas at Austin, Marine Science Institute.

Chamberlain, G. W., R. J. Miget, and M. G. Haby, eds. 1987. *Manual on red drum aquaculture*. Corpus Christi, TX: Texas Agricultural Extension Service, Texas A&M University.

Chamberlain, G. W., R. J. Miget, and M. G. Haby, eds. 1990. *Red drum aquaculture*. TAMU-SG-90-603. Galveston: Texas A&M University Sea Grant Program.

Colura, R. L., A. Henderson-Arzapalo, and A. F. Maciorowski. 1991. Culture of red drum. In *CRC Handbook of mariculture, Vol. II: Finfish aquaculture*, ed. J. P. McVey, pp. 149–66. Boca Raton: CRC Press.

Overstreet, R. M. 1983. Aspects of the biology of the red drum, *Sciaenops ocellatus*, in Mississippi. *Gulf Research Reports, Supplement* 1:45–68.

Eel

SCIENTIFIC AND COMMON NAMES: ***Anguilla rostrata,* American eel**

DESCRIPTION AND DISTINCTIVE CHARACTERISTICS: Elongate fish with tiny scales; caudal fin extends from the center of the back around the tail and forward on the belly with no separation into dorsal, caudal, and ventral portions. Eels are dark brown or olive green on the dorsal surface and white to golden underneath. Upon reaching maturity, eels develop an all-over silver color.

HABITAT RANGE: The geographic range of the American eel extends from Greenland to the Gulf of Mexico. Males tend to remain in estuaries, but females journey far upstream into lakes and headwaters of rivers. The major portion of their lives are spent in fresh or brackish water.

REPRODUCTION AND DEVELOPMENT: Eels spawn in the Sargasso Sea hundreds of miles offshore. Each female lays 1–20 million eggs at a depth of about 450 ft, usually during January–March. Eggs float to the surface and hatch into tiny prelarvae approximately 1 mm in length. They soon metamorphose into transparent, ribbon-shaped larvae called *leptocephali,* weak swimmers that drift with the ocean currents, moving up freshwater streams along the coast.

AGE AND GROWTH: Females live 10–13 years in fresh water and commonly reach a length of 3 ft; lengths of 5 ft are rare. Males are much smaller, rarely longer than 20 in. Weights of 0.5–2 lb are common for eels 15–130 in. long.

FOOD AND FEEDING: In captivity small earthworms, shrimps, tubifex worms, and daphnia are used to get eels to start feeding. Commercial feeds compounded for eels, consisting largely of fish meal, are available.

PARASITES AND DISEASE: Bacterial diseases (red fin disease, red spot disease, gill rot, tail rot); fungus (cotton covered disease); protozoans ("Ich" disease, *Trichodina* gill infection, suctorian gill disease, skin disease caused by myxosporean infection, beko disease caused by microsporean infection); monogenetic trematodes, *Gyrodactylus* and *Dactylogryus;* copepods (anchor worms);

nematode, anguillicola disease. Diseases for which the cause is unknown: connected hole disease, abdominal dropsy disease, and gill nephritis.

PREDATORS AND COMPETITORS: Plankton feeders are predatory on early larval stages; large fish on juveniles and adults.

AQUACULTURAL POTENTIAL: Because their life cycle is not closed (i.e., they cannot be spawned in captivity at this time), raising eels depends on capturing wild *elvers* (young) as they migrate upstream. This procedure works well in Japan. There are strong markets in Europe and Japan (European eel, *Anguilla anguilla*). Disease prevention and control are important aspects of eel culture, as there have been considerable losses. Diseases have been spread by the shipment of diseased elvers: In 1969, European elvers shipped from France to Japan were blamed for an outbreak of ichthyopthiriasis.

REGIONS WHERE FARMED AND/OR RESEARCHED: Florida; France and Japan.

REFERENCES

Angel, N. B., and W. R. Jones. 1974. *Aquaculture of the American eel (*Anguilla rostrata*)*. Raleigh: Industrial Extension Service, School of Engineering, North Carolina State University.

Deelder, C. L. 1984. *Synopsis of biological data on the eel,* Anguilla anguilla *(Linnaeus, 1758)*. FAO Fisheries Synopsis no. 80, rev. 1. Rome: Food and Agriculture Organization of the United Nations.

Rickards, W. L., W. R. Jones, and J. E. Foster. 1978. Techniques for culturing the American eel. *Proceedings of the World Mariculture Society* 9:641–6.

Usui, A. 1974. *Eel culture*. West Byfleet (U.K): Fishing News Books.

Grouper

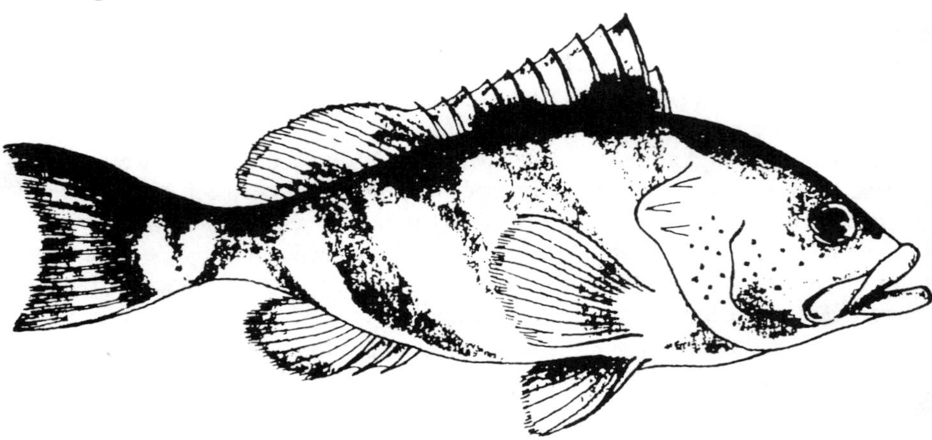

SCIENTIFIC AND COMMON NAMES: ***Epinephelus morio,* Red grouper**
Related: black grouper, *Mycteroperca bonaci.*

DESCRIPTION AND DISTINCTIVE CHARACTERISTICS: The red grouper has small scales, large eyes, and a robust body. Coloration is highly variable, but generally uniformly brownish-red with a lighter ventral coloration and a transient pattern of whitish spots; it may be changed to blend in with surroundings. The red grouper is distinguished from other groupers by the second (longest) spine on the dorsal fin, and by the interspinous membrane, which is not notched.

HABITAT RANGE: Generally on rocky or muddy bottoms. Younger fish tend to inhabit the inshore waters. Groupers are found at depths of up to 625 ft in the temperature range of 59–86 °F.

REPRODUCTION AND DEVELOPMENT: Some groupers are hermaphrodites, as is the red grouper, which changes from female to male after about 5–10 years of age. Females become mature at age 4–6, and individuals produce 200,000–700,000 eggs. Spawning occurs during late April–early May, usually when the water temperature reaches about 25 °F. Larvae metamorphose to juveniles after about a month.

AGE AND GROWTH: Grouper may live as long as 30 years. They grow rapidly in cages, reaching about 21 oz in six months or so. The rate of growth for each sex is about the same, but males reach a slightly longer ultimate length than do females.

FOOD AND FEEDING: Opportunistic carnivores, feeding on a variety of cephalopods, crustaceans, and other invertebrates, as well as on fishes.

PARASITES AND DISEASE: Copepods cause skin lesions; fungal infections; reddening of the caudal peduncle, perhaps caused by a vibriolike infection; fatty degeneration of the liver accompanied by hemorrhaging of the muscles and eyes; swim bladder inflammation; digenetic trematodes (3 spp.), cestodes, and nematodes.

PREDATORS AND COMPETITORS: May be preyed upon by other groupers and sharks. Compete with other groupers for food because of habitat overlap; may also compete with jacks, snappers, barracudas, and sharks.

AQUACULTURAL POTENTIAL: Very popular and expensive fish with many attributes good for aquaculture: They grow rapidly, can be induced to spawn, and are hardy. In controlling reproduction an important problem with some species is that they are protogynous hermaphrodites. Very large fish that have become males must be found and maintained for spawning purposes. Groupers are commercially cultured in cages and ponds in many areas of the Far East; trials in the Atlantic have all been of an experimental nature. Disease problems and cost of feeding may hinder commercial development.

REGIONS WHERE FARMED AND/OR RESEARCHED: A variety of grouper species are farmed in Malaysia, Thailand, Singapore, and Hong Kong; most culture facilities are small scale. Farming trials have been made in Kuwait. Research is under way at the University of Miami, RSMAS; the Florida Department of Natural Resources; the Mote Marine Laboratory, Florida; and in the Cayman Islands, Cuba, Venezuela, and Mexico.

REFERENCES

Jory, D. E., and E. S. Iversen. 1989. *Species profiles: Life histories and environmental requirements of coastal fishes and invertebrates (South Florida). Black, red and Nassau groupers.* Biological Report 82(11.110). Slidell, LA: National Wetlands Research Center, U. S. Fish and Wildlife Service.

Tucker, J. W., Jr. and D. E. Jory. 1991. Marine fish culture in the Caribbean region. *World Aquaculture* 22(1):10–27.

Halibut

NEWLY HATCHED LARVA (Stage 1)

Showing prominent yolk sac.

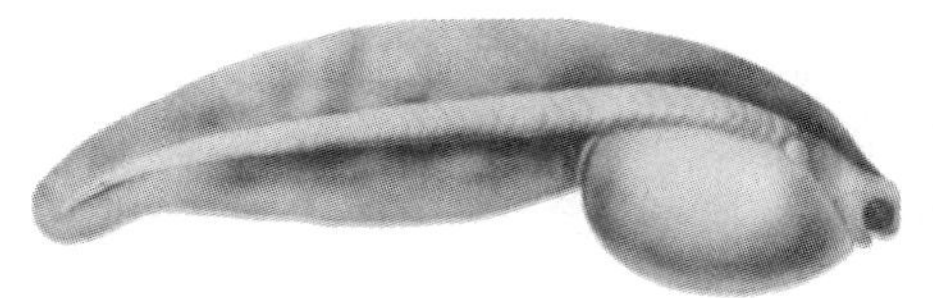

Approximately 9 mm in length.

POSTLARVA (Stage 3)

Yolk sac has been absorbed.

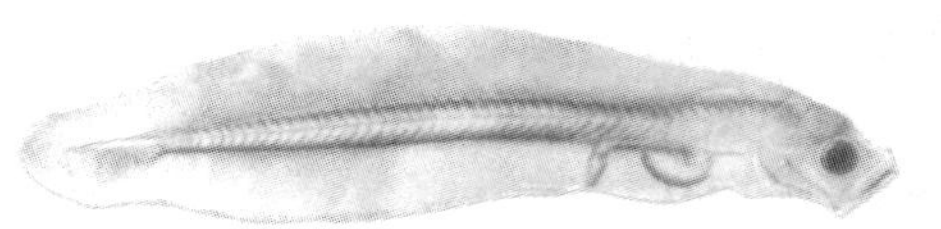

Approximately 16 mm in length.

POSTLARVA (Stage 7)

Approximately 21 mm in length.

POSTLARVA (Stage 9)

Showing the beginning of eye migration.

Approximately 25 mm in length.

YOUNG HALIBUT

Adapted to bottom life.

Approximately 35 mm in length.

SCIENTIFIC AND COMMON NAMES: ***Hippoglossus hippoglossus*, Atlantic halibut**

Also called cherry belly and gray.

DESCRIPTION AND DISTINCTIVE CHARACTERISTICS: The largest member of the family Pleuronectidae of flounders. The Atlantic halibut is "righteye" or right-sided; that is, the color and the eyes both move to the right side when the fish takes up a bottom-dwelling existence.

HABITAT RANGE: Southward on the coast of Europe to France, and on the eastern seaboard of North America from New Jersey northward. It is a cold-water fish, living in water above 37 °F and in close association with the bottom.

REPRODUCTION AND DEVELOPMENT: Off Canada's Atlantic Coast, this flatfish spawns in deep water during late winter and early spring. A large female may release over two million eggs during multiple spawnings in one season. The pelagic larvae are able to feed (i.e., eyes, jaws, and teeth function) about one month after hatching; two months thereafter, they metamorphose into juveniles.

AGE AND GROWTH: Most range 4–6 ft in length and weigh 50–200 lb. A 9-ft specimen weighing 625 lb when dressed was caught off Massachusetts. Market size of 11–33 lb is reached in three to five years. In general, growth is slow.

FOOD AND FEEDING: Eat invertebrates, including crabs, lobsters, clams, and mussels; but the chief item in the diet is usually finfish.

PARASITES AND DISEASE: Some 34 parasite species, internal and external, have been found in the Atlantic halibut, including trematodes, cestodes, nematodes, acanthocephalans, and copepods. Some internal parasites infect the blood.

PREDATORS AND COMPETITORS: As young, they serve as food for many fishes. Sharks and seals are believed to prey on large halibut.

AQUACULTURAL POTENTIAL: A very desirable market fish. Artificial fertilization of halibut eggs was carried out during the early 1930s. This fish is now considered ready for large-scale commercial production despite unsolved problems of feeding live food and diseases caused by microbes: Recent trials in Norway gave larval survival rates as low as 1 percent. Also, benthic algae in the "green water" larviculture technique become entangled in halibut larvae and can cause high mortality. The lack of competent personnel to staff hatcheries may delay commercial production. Efficient grow-out technology and feeds to increase growth rates have yet to developed.

REGIONS WHERE FARMED AND/OR RESEARCHED: Norway since 1974, with greater emphasis starting in 1983; also the United Kingdom and Canada.

REFERENCES

Blaxter, J. H. S., D. Danielssen, E. Moksness, and V. Oiestad. 1983. Description of the early development of the halibut *Hippoglossus hippoglossus* and attempts to rear the larvae past the first feeding. *Marine Biology* 73:99–107.

Bolla, S., and I. Holmefjord. 1988. Effect of temperature and light on development of Atlantic halibut larvae. *Aquaculture* 74:355–8.

Haug, T. 1990. Biology of the Atlantic halibut, *Hippoglossus hippoglossus* (L. 1758). *Advances in Marine Biology* 26:1–70.

Scott, W. B., and M. G. Scott. 1988. *Atlantic fishes of Canada.* Canadian Bulletin of Fisheries and Aquatic Sciences no. 219. Toronto: University of Toronto Press.

Milkfish

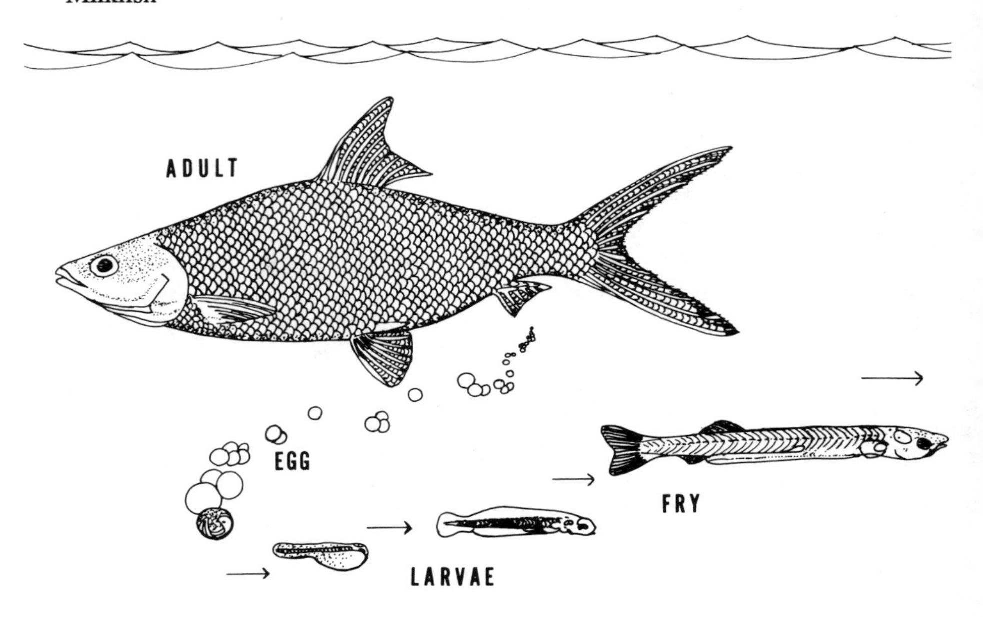

SCIENTIFIC AND COMMON NAMES: ***Chanos chanos*, Milkfish**
Called *bandeng* in Indonesia, *bangos* in the Philippines, and *awa* in Hawaii.

DESCRIPTION AND DISTINCTIVE CHARACTERISTICS: The body is elongate and slightly compressed, the head pointed, and tail deeply forked. Eyes are large and the mouth is weak. The fish is bright silvery or bluish above, with a golden luster along the sides of some specimens reported. There is a yellowish cast to the rather strong dorsal and pectoral fins.

HABITAT RANGE: From longitude 40° E to about 100° W and from latitude 30–40° N. to 30–40° S. Milkfish occur in the tropical and subtropical areas of the Indian and Pacific Oceans. The species has not been reported from either the Atlantic or Pacific coast of Central or South America.

REPRODUCTION AND DEVELOPMENT: Maturation is reached rather late in life: at least age 4 years in males and 5 or older in females. Fertilization is external, the eggs being released in the open sea near shore to float to the surface. As many as seven million eggs have been estimated to be produced by a single large female. All attempts to obtain eggs from females either by natural spawning or by stripping in ponds have failed. Milkfish spawn at different times in various parts of their wide geographic distribution. In some waters they spawn twice in one year, late in the spring and during the fall; in others they spawn only during the summer.

AGE AND GROWTH: The growth of milkfish is easily adjusted by the amount of food provided to them, and by the stocking rates in ponds. They may reach as much as 1.8 lb after a year and 3.3–4.4 lb at the end of two. At about age 4, *Chanos* may reach a weight of 5.5–7.7 lb in ponds.

FOOD AND FEEDING: Milkfish are herbivores and eat diatoms, blue-green algae, filamentous algae, and occasionally some animals, such as nematodes and copepods. They feed throughout the year either at the surface or on the bottom of ponds.

PARASITES AND DISEASE: Parasites include viruses, bacteria (*Vibrio,* red spot disease), fungi, isopods, digenetic trematodes, and copepods. Miscellaneous noninfectious diseases: gas bubble disease and stress-induced discoloration.

PREDATORS AND COMPETITORS: In nursery and grow-out ponds, young fish are eaten by frogs, snakes, and lizards. Larger milkfish are eaten by crocodiles.

AQUACULTURAL POTENTIAL: This is an extremely important aquacultural species wherever it occurs. Sexual maturity is reached late in life and it is difficult to induce mature fish to spawn in captivity. Because large-scale artificial propagation has not been successful, seed fish are still obtained from natural reproduction. Larval rearing in outdoor ponds is successful. Records of pond-reared milkfish in Hawaii go back 300 years, when many ponds were located along the coast; today the few milkfish ponds still rely on wild-caught fish for stocking. In recent years the price and demand for milkfish have dropped.

REGIONS WHERE FARMED AND/OR RESEARCHED: Indonesia, the Philippines, Taiwan, and Hawaii. The species is receiving considerable research attention at the Oceanic Institute in Honolulu, primarily on maturation and spawning of brood stock.

REFERENCES

Benedicto-Dormitorio, A., ed. 1983. *A compilation of SEAFDEC AQD technical papers on milkfish & other finfishes.* 2 vols. Tigbauan (Philippines): Aquaculture Dept., Southeastern Asian Fisheries Development Center.

Kelley, C., and C. S. Lee. 1991. Milkfish culture in the U.S. In *CRC handbook of mariculture, Vol. II: Finfish aquaculture,* ed. J. P. McVey, pp. 211–26. Boca Raton, FL: CRC Press.

Lee, C. S., M. S. Gordon, and W. O. Watanabe, eds. 1986. *Aquaculture of milkfish* (Chanos chanos): *State of the art.* Waimanalo, Hawaii: Oceanic Institute.

Schuster, W. H. 1960. *Synopsis of biological data on milkfish* Chanos chanos *(Forsskal, 1775).* FAO Fisheries Biology Synopsis no. 4. Rome: Food and Agriculture Organization of the United Nations.

Smith, I. R., and K. C. Chong. 1984. Southeast Asian milkfish culture: Economic status and prospects. In *Advances in milkfish biology and culture; proceedings of the 2nd International Milkfish Aquaculture Conference, 1983,* ed. J. V. Juario, R. P. Ferraris, and L. V. Benitez, pp. 1–20. Manila: Island Publishing House.

Mullet

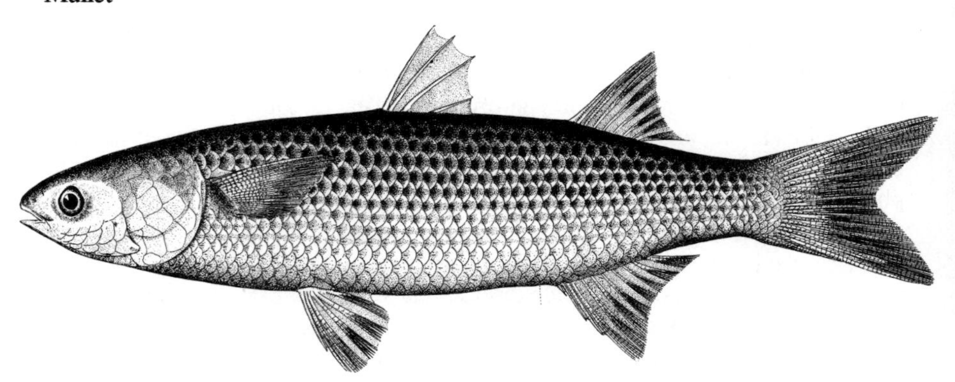

SCIENTIFIC AND COMMON NAMES: ***Mugil cephalus*, Striped mullet**

DESCRIPTION AND DISTINCTIVE CHARACTERISTICS: The body is elongate, grayish above and silvery on the sides and below; there is no lateral line. There are two dorsal fins, the first having four sharp spines. Eyes have gelatinous lids anterior and posterior to pupil.

HABITAT RANGE: Found nearly worldwide in temperate, tropical, and subtropical, mainly marine and brackish, waters; occasionally found in fresh water.

REPRODUCTION AND DEVELOPMENT: In the Philippines, striped mullet spawn in the open sea during the dry season, and the fry appear in shallow coastal waters and around the mouths of rivers in April–July. In the U.S. Southeast, spawning occurs during October–February in offshore waters. Fecundity varies widely by size of female and geographic location. About three million large eggs are released by one female in a given spawning.

AGE AND GROWTH: In Japan, these fish grow to about 7 oz between March–April and November–December. In India they reach about 18 in. in one year, whereas in Florida they reach 12 in. in three years.

FOOD AND FEEDING: Striped mullet eat algae growing on the substrate, decayed organic matter, and small crustaceans; young mullet eat plankton and algae in the water column.

PARASITES AND DISEASE: Parasites include protozoans, monogeneans, digeneans, cestodes, nematodes, acanthocephalans, copepods, branchiurans, isopods, and leeches.

PREDATORS AND COMPETITORS: Predatory fish and birds.

AQUACULTURAL POTENTIAL: Since mullet feed on algae, diatoms, small crustaceans, and decayed organic matter, there is little need to feed them, thus eliminating feed and labor costs. They can be raised in combination with other

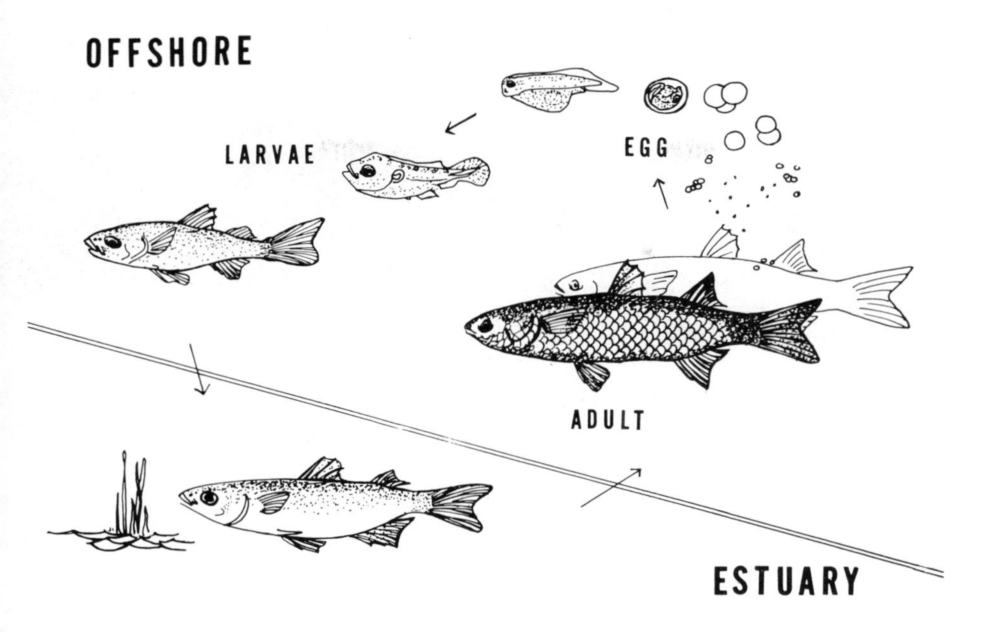

Mullet life cycle.

herbivorous fish and shrimp and thereby increase the productivity of an aquacultural operation. However, the price for mullet is quite low in many areas, despite the desirable taste of these fish, thus making mullet farming unprofitable. Also, there have been difficulties associated with artificial spawning and rearing of larvae.

REGIONS WHERE FARMED AND/OR RESEARCHED: China, Korea, the Philippines, Hawaii, Israel.

REFERENCES

Lee, C. S., and C. D. Kelley. 1991. Artificial propagation of mullet in the United States. In *CRC Handbook of mariculture, Vol. II: Finfish aquaculture,* ed. J. P. McVey, pp. 193–209. Boca Raton, FL: CRC Press.

Nash, C. E., and Z. H. Shehadeh, eds. 1980. *Review of breeding and propagation techniques for grey mullet,* Mugil cephalus *L.* ICLARM Studies and Reviews, no. 3. Manila (Philippines): International Center for Living Aquatic Resources Management.

Oren, O. H., ed. 1981. *Aquaculture of grey mullets.* International Biological Programme, no. 26. New York: Cambridge University Press.

Muskellunge

SCIENTIFIC AND COMMON NAMES: ***Esox masquinongy*, Muskellunge**

DESCRIPTION AND DISTINCTIVE CHARACTERISTICS: A powerful fish with an elongated body and head, and a single dorsal fin located toward the posterior end; all fins are soft rayed. The mouth is large and has long jaws. There are dark vertical bars on both sides of the body. It is closely related to the pike.

HABITAT RANGE: Found in fresh water, in the Great Lakes and U.S.-side St. Lawrence River; also the upper Mississippi River drainage as far west as Minnesota.

REPRODUCTION AND DEVELOPMENT: In early spring the females lay as many as 100,000 eggs, which adhere to weeds and the bottom; about two weeks later the fry hatch.

AGE AND GROWTH: Fry and young juveniles grow very rapidly, and may reach 6 in. in 60 days. Most fish weigh about 15 lb; large ones reach over 60 in. and weigh 70 lb.

FOOD AND FEEDING: Fry, a few days old, eat zooplankton; they then switch to large quantities of forage fish. Adults make serious inroads into stocks of such fish as suckers, perch, and carp, but eat practically any fish that gets too close. Sometimes called "freshwater barracuda" because of the similarity of the two fishes, their ways of stalking prey, and their highly predatory habits, they are generally voracious feeders.

PARASITES AND DISEASE: Numerous species of protozoans, trematodes, cestodes, nematodes, acanthocephalans, mollusks, and crustaceans.

PREDATORS AND COMPETITORS: Cannibalistic; young are eaten by a variety of predacious fishes.

AQUACULTURAL POTENTIAL: The muskellunge has been artificially raised for over a century in New York and Wisconsin. The high cost of using live feeds and cannibalistic behavior during rearing have discouraged private aquaculture. Most muskellunge culture programs are government sponsored and directed toward augmenting wild stocks for recreational fishing. Fingerling releases have been used to compensate for the inroads anglers make in wild populations. There has been variable success in raising swimming fry to fingerlings of about 3 in. long. Hybrid muskellunge (northern pike × muskellunge) were developed in the 1930s and could be grown on commercially available artificial feeds; this was a big step forward. A number of problems remain to be addressed before survival of early stages can be predictable.

REGIONS WHERE FARMED AND/OR RESEARCHED: Kentucky, Ohio, Michigan, Minnesota, Missouri, Pennsylvania, Iowa, and Wisconsin; Ontario, Canada.

REFERENCES

Hall, G. E., ed. 1986. *Managing muskies: A treatise on the biology and propagation of muskellunge in North America; proceedings of an international symposium, 1984*. American Fisheries Society Special Publication no. 15. Bethesda, MD: American Fisheries Society.

Stickney, R. R., ed. 1986. *Culture of nonsalmonid freshwater fishes*. Boca Raton, FL: CRC Press.

Paddlefish

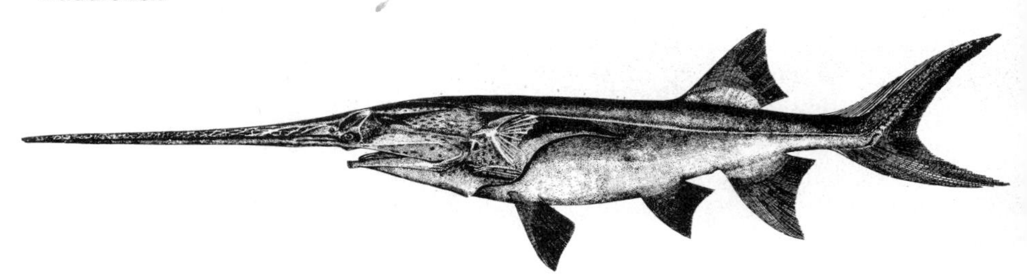

SCIENTIFIC AND COMMON NAMES: ***Polyodon spathula,* Paddlefish (U.S.)**
Related: *Psephurus gladius* (China).

DESCRIPTION AND DISTINCTIVE CHARACTERISTICS: A primitive fish with a predominantly cartilaginous skeleton and a strongly developed, spoonlike or paddlelike snout. Paddlefish have smooth skins and a large subterminal mouth. They reach up to 6 ft in length and can weigh more than 200 lb.

HABITAT RANGE: Muddy waters in the Mississippi Valley; *P. gladius* in the Yangtze River in China.

REPRODUCTION AND DEVELOPMENT: Male paddlefish reach sexual maturity in 7–9 years, females in 10–12. Spawning is triggered at temperatures of about 60 °F. Adhesive eggs attach to clean gravel in areas of rapid water flow.

AGE AND GROWTH: Growth is rather rapid, especially in first year of life (up to 20 in. total length). Paddlefish may live to about 30 years. Weight increases rapidly after about age 5.

FOOD AND FEEDING: Paddlefish are filter feeders, consuming mostly insect larvae and zooplankton. Food is captured while paddlefish swim about with their unusually large mouth held open. The heavy "paddle" may allow the species to maintain its vertical position in the water column by compensating for the drag caused by the large mouth.

PARASITES AND DISEASE: Protozoans, cestodes, and trematodes. Eggs are parasitized by a coelenterate (*Polypodium* sp.). (See Dillard et al. 1986.)

PREDATORS AND COMPETITORS: Fry are eaten by predacious water insects.

AQUACULTURAL POTENTIAL: Rearing techniques for paddlefish were developed in Missouri about 40 years ago. They can be reared in polyculture, usually with channel catfish. In recent years culturing techniques have improved, and a strong and increasing market for roe for caviar has developed; the same cannot be said for paddlefish flesh, though it is occasionally smoked to improve the flavor. Only a few commercial growers are producing paddlefish, in low volume and at high risk. Their numbers may increase if markets for the flesh can be enlarged, if culture techniques can become more widely known, and if wild stocks continue to be depleted by fishing. The make-or-break aspect of paddlefish farming is in the markets and marketing strategies used.

REGIONS WHERE FARMED AND/OR RESEARCHED: Auburn University, Alabama; Iowa, Missouri, Montana, South Dakota, and Tennessee.

REFERENCE

Dillard, J. G., L. K. Graham, and T. H. Russell. 1986. *The paddlefish: Status, management and propagation*. North Central Division, American Fisheries Society, Special Publication no. 7. Columbus, OH: North Central Division, American Fisheries Society.

Perch

SCIENTIFIC AND COMMON NAMES: ***Perca flavescens,* Yellow perch** Also called common perch, ringed perch, raccoon perch, striped perch.

DESCRIPTION AND DISTINCTIVE CHARACTERISTICS: Body rather rounded; the dorsal fins are separated with strong spines. The dorsal body surface is dark, and there are seven or so vertical bars on the sides.

HABITAT RANGE: In fresh water, from Nova Scotia to South Carolina; apparently prefer quiet streams and ponds and slow-moving rivers. Yellow perch are found in the Hudson Bay drainage system of eastern Canada south to Kansas and northern Missouri, Illinois, Indiana, and western Pennsylvania. They have also been introduced to the western United States.

REPRODUCTION AND DEVELOPMENT: Eggs are laid at night during the spring, in a gelatinous rope that may stretch 7 ft long, and become entwined in aquatic weeds. Fry hatch in about a week; active swimming may occur within two days after hatching.

AGE AND GROWTH: Average weight is 1–2 lb, maximum about 4. May reach a foot in length. Can reach a market size in two or three years.

FOOD AND FEEDING: Larvae feed on zooplankton; juveniles and adults eat small worms, crustaceans, insects, and fishes. It is apparently economically impossible to raise yellow perch to the commercial market size using live food; they can, however, be trained to accept artificial diets.

PARASITES AND DISEASE: Parasites include protozoans, trematodes (numerous spp.), cestodes, nematodes, acanthocephalans, leeches, mollusks, and crustaceans.

PREDATORS AND COMPETITORS: Preyed on by a variety of fishes when small; pike and pickerel eat larger, adult yellow perch.

AQUACULTURAL POTENTIAL: Yellow perch is a valuable commercial and recreational food and forage fish, and has been cultured since the early 1900s for stocking public and private lakes. However, the margin between the cost of raising yellow perch using current culture systems and the market price is too small to encourage commercial culture.

REGIONS WHERE FARMED AND/OR RESEARCHED: Michigan, Wisconsin, and California.

REFERENCE

Stickney, R. R., ed. 1986. *Culture of nonsalmonid freshwater fishes.* Boca Raton, FL: CRC Press.

Pike

SCIENTIFIC AND COMMON NAMES: ***Esox lucius,*** **Northern pike** Also called great northern pike, pickerel.

DESCRIPTION AND DISTINCTIVE CHARACTERISTICS: Closely related to the muskellunge, to which it is bodily very similar; these two fishes are known to cross-fertilize. The robust body is spotted, with pale oblong spots on a gray-green background. The dorsal fin is located far back on the body, and the large, powerful mouth has razor-sharp teeth.

HABITAT RANGE: In warm-water streams and shallow weedy lakes, from Alaska south to Labrador, New England, and the Ohio Valley; transplanted to the Pacific Northwest, Europe, and Asia.

REPRODUCTION AND DEVELOPMENT: Pike may ascend small streams and lay eggs around the margins of lakes. A large female may release over 100,000 eggs during a spawning season (spring).

AGE AND GROWTH: Known to reach 8–12 in. at the end of the first summer; pike about age 9 were 36 in. Some reach 4.5 ft in length and weigh 45 lb—the record is 100 lb and 68 in.—but their usual weight is about 3 lb.

FOOD AND FEEDING: Feeds on a variety of fishes and is cannibalistic. During the first week or two, pike feed on small insects and crustaceans, but they rapidly change to a diet of small fish fry available at the time.

PARASITES AND DISEASE: Pike, over a range of sizes and geographic locations, are hosts to parasites from many phyla, including protozoans, trematodes, nematodes, acanthocephalans, crustaceans, hirudineans, and fungi, with numerous genera and species in each. Many of these phyla contain species known to cause high mortality in pike. Viral diseases include pike fry rhabdovirus disease, Egtved disease, lymphocystis disease, and lymphosarcoma. Bacterial diseases reported are red sore disease, black-spot disease, furunculosis, and mycobacteriosis.

PREDATORS AND COMPETITORS: Minnows eat young pike, and pike cannibalize their larvae.

AQUACULTURAL POTENTIAL: Pike is an excellent tasting popular food fish and easy to propagate artificially. Hatchery production is used for stocking natural waters for sport and commercial fisheries. Sale to commercial markets is rather limited.

REGIONS WHERE FARMED AND/OR RESEARCHED: Stocking pike for recreational and commercial fishing is practiced in many countries throughout its range. Pike releases have been made in 26 reservoirs in nine U.S. states, and Colorado and Wisconsin have stocking programs.

REFERENCES

Crossman, E. J., and J. M. Casselman. 1987. *An annotated bibliography of the pike,* Esox lucius *(Osteichthyes: Salmoniformes).* Toronto: Royal Ontario Museum.

Raat, A. J. P. 1988. *Synopsis of biological data on the northern pike,* Esox lucius *Linnaeus, 1758.* FAO Fisheries Synopsis no. 30, rev. 2. Rome: Food and Agriculture Organization of the United Nations.

Stickney, R. R., ed. 1986. *Culture of nonsalmonid freshwater fishes.* Boca Raton, FL: CRC Press.

Pompano

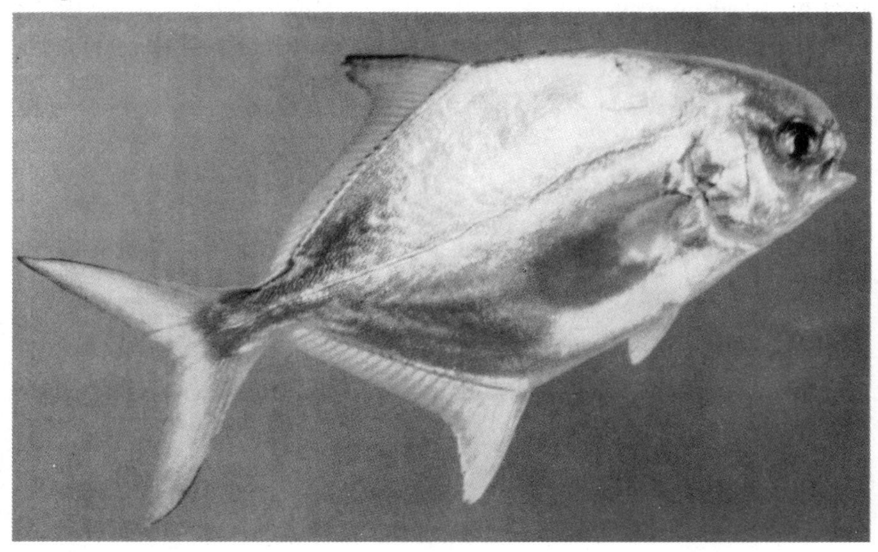

SCIENTIFIC AND COMMON NAMES: ***Trachinotus carolinus,*** **Florida pompano**

DESCRIPTION AND DISTINCTIVE CHARACTERISTICS: A thin, deep-bodied fish of the family of jacks (Carangidae), with a silvery body, shading to metallic blue above and to a golden yellow ventrally. The upper fins are dark, whereas the lower fins are yellow, shaded with blue. The caudal and dorsal fins are deeply forked; spines and soft rays are found in the dorsal and anal fins. A blue line appears above and in front of the eyes, and the lower jaw protrudes.

HABITAT RANGE: This migratory species is found along the southeastern coast of the United States and in the Gulf of Mexico. Young live in the upper reaches of the shallow coastal waters.

REPRODUCTION AND DEVELOPMENT: Spawning takes place offshore in areas where ocean currents will carry the young back to shore.

AGE AND GROWTH: Florida pompano are believed to grow about 1 in. monthly. They range in weight from about 1.5 to 3 lb, but may reach about 6 lb.

FOOD AND FEEDING: Feed on bottom organisms, mainly clams, crabs, shrimp and mussels.

PARASITES AND DISEASE: Parasites include protozoans, cestodes, monogeneans, digeneans, and crustaceans.

PREDATORS AND COMPETITORS: Large fish are predators.

AQUACULTURAL POTENTIAL: The strong market demand and high price received for this species suggests a strong candidate for aquaculture. Most of the life history has been researched, but U.S. farming trials have not been successful. Food conversion has been rather poor once the fish reach a weight of about 0.5 lb; perhaps with an improved feed for this species, profitable farming could be achieved.

In Alabama, polyculture (pompano with white shrimp [*Penaeus vannamei* and *P. stylirostris*]) suggested a profitable operation, whereas pompano monoculture was uneconomical.

REGIONS WHERE FARMED AND/OR RESEARCHED: Florida Keys; Alabama Marine Resources Division, Gulf Shores; Venezuela and the Dominican Republic.

REFERENCES

Berry, F., and E. S. Iversen. 1966 Pompano: Biology fisheries and farming potential. *Proceedings of the Gulf and Caribbean Fisheries Institute* 19:116–28.

Fischer, W., ed. 1978. *FAO species identification sheets for fishery purposes, Western Central Atlantic (Fishing Area 31)*, vol. 2. Rome: Food and Agriculture Organization of the United Nations.

Gomez, A. 1984. Crecimiento del pampano *Trachinotus carolinus* (Linnaeus) vacunados contra la vibriosis en la Isla de Margarita, Venezuela. *Memorias de la Asociacion Latinoamericana de Acuicultura* 5(3):703–8.

Iversen, E. S., and F. H. Berry. 1969. Fish mariculture: Progress and potential. *Proceedings of the Gulf and Caribbean Fisheries Institute* 21:163–76.

McMaster, M. F. 1988. Pompano aquaculture: Past success and present opportunities. *Aquaculture Magazine* 14(3): 28–34.

Smith, T. I. J. 1973. *The commercial feasibility of rearing pompano,* Trachinotus carolinus *(Linnaeus) in cages.* Florida Sea Grant Technical Bulletin no. 26. Coral Gables, FL: University of Miami Sea Grant College Program.

Tatum, W. M. 1973. Comparative growth, mortality and efficiency of pompano (*Trachinotus carolinus*) receiving a diet of ground industrial fish with those receiving a diet of trout chow. *Proceedings of the 3rd annual World Mariculture Society Workshop*, 1972:65–74.

Trimble, W. C. 1980. Production trials for monoculture and polyculture of white shrimp (*Penaeus vannamei*) or blue shrimp (*P. stylirostris*) with Florida pompano (*Trachinotus carolinus*) in Alabama, 1978–1979. *Proceedings of the World Mariculture Society* 11:44–59.

Pumpkinseed

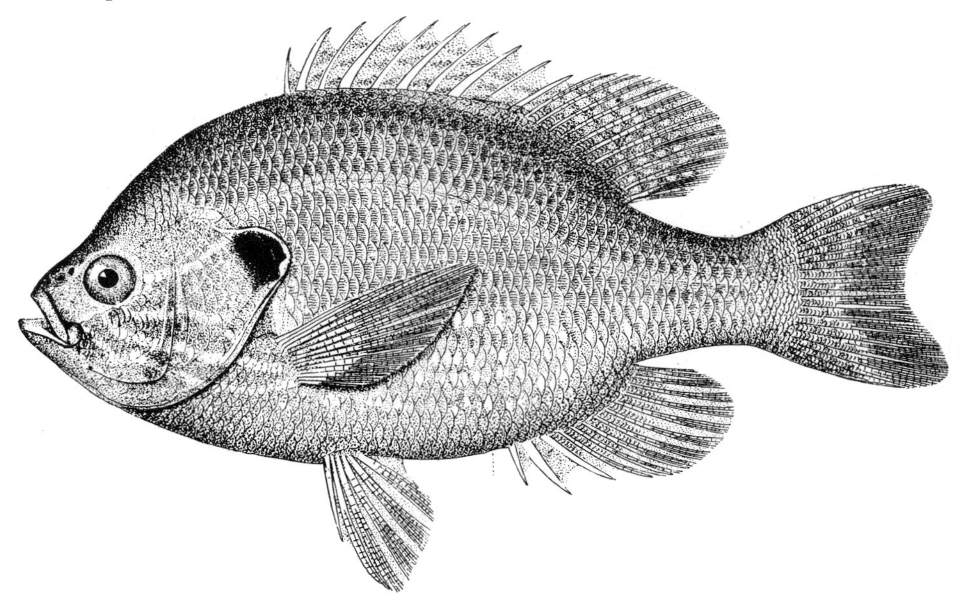

COMMON AND SCIENTIFIC NAME: ***Lepomis gibbosus*, Pumpkinseed** Also called common sunfish, pond perch, yellowbelly, sunny (among others). Green sunfish (*L. cyanellus*) have been crossed with pumpkinseeds to produce a larger fish.

DESCRIPTION AND DISTINCTIVE CHARACTERISTICS: Its body is laterally compressed, the mouth is small, and the back rather humped. The pectoral fins are long and pointed; the dorsal and anal fins have spines in advance of the soft rays. A good character used to identify the pumpkinseed is the bright orange spot on the long opercular lobe. The body may be covered with orange spots, and vertical bars may be present. Young fish have yellow bellies.

HABITAT RANGE: Fresh water in lakes and streams. Covers a wide north–south range in the United States, extending from the Gulf States' north waters to southern Canada, and the Maritime Provinces south to Florida. Common in the Great Lakes region.

REPRODUCTION AND DEVELOPMENT: Sexual maturity is reached early, usually by age 1 or 2 years. Spawning takes place in the spring and summer in shallow water.

AGE AND GROWTH: Attain a length of about 12 in., and reach an age of 7 years. In overpopulated lakes, adults may commonly be only about 4 in. long.

FOOD AND FEEDING: Eat insects, crayfish, crustaceans, and snails; may be cannibalistic.

PARASITES AND DISEASE: Protozoans, trematodes (numerous species), cestodes, nematodes, acanthocephalans, leeches, mollusks, and crustaceans.

PREDATORS AND COMPETITORS: Preyed upon by basses. Pumpkinseeds tend to overreproduce, which can cause stunting of population in ponds and tanks.

AQUACULTURAL POTENTIAL: Although a popular fish, pumpkinseeds are small and hence not very desirable as a farmed species for food. The value of sunfish as a forage fish for the popular largemouth and smallmouth basses has been known since the 1930s. Their high reproduction rate is a desirable feature in hatcheries where predatory sportfish are being reared. Stocking hybrid sunfish as a forage fish is not a desirable management practice because the hybrids have a low reproductive potential resulting in so few offspring that a healthy bass population cannot be sustained.

REGIONS WHERE FARMED AND/OR RESEARCHED: North Carolina, Oregon, Arizona, Pennsylvania, Tennessee, and Texas.

REFERENCES

McLarney, W. 1984. *The freshwater aquaculture book; a handbook for small scale fish culture in North America.* Point Roberts, WA: Hartley & Marks.

Stickney, R. R., ed. 1986. *Culture of nonsalmonid freshwater fishes.* Boca Raton, FL: CRC Press.

Salmon

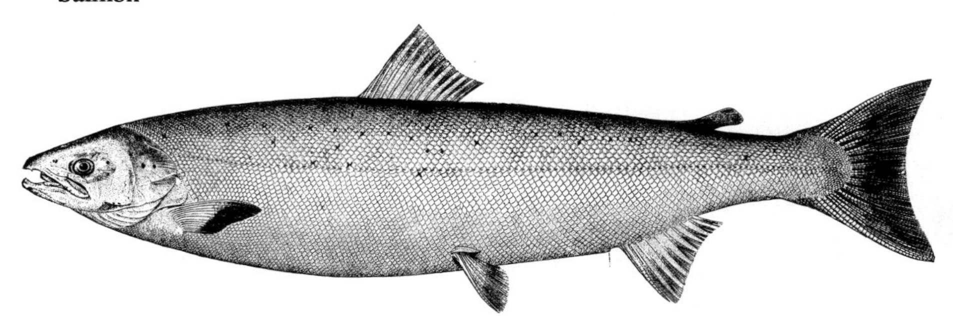

SCIENTIFIC AND COMMON NAMES: ***Salmo salar*, Atlantic salmon**

DESCRIPTION AND DISTINCTIVE CHARACTERISTICS: A long-bodied fish with a naked head. Fins are spineless, and there is a well-developed adipose fin. During their freshwater existence, Atlantic salmon are marked with vertical bars and X-shaped black spots; but upon entering salt water, they develop an overall silver color.

HABITAT RANGE: Both sides of the Atlantic, Greenland to Cape Cod; fresh water when young, salt water as juveniles and adults.

REPRODUCTION AND DEVELOPMENT: Adults return to rivers to spawn (usually November–December); females lay about 5,000 eggs in gravel nests. Fertilization is external. Eggs take three months to hatch; young stay in fresh water until they reach about 6 in. in length. Unlike most Pacific salmon (genus *Oncorhynchus*), Atlantic salmon do not die after the initial spawning.

AGE AND GROWTH: Atlantic salmon spend two or three years in fresh water and then one or two (but up to about five) years in salt water. Average weight upon return is roughly 15 lb; length is usually 1.5–3 ft. Record catches have been made of Atlantic salmon 5 ft long and weighing 84 lb.

FOOD AND FEEDING: Highly carnivorous, with a diet that includes sand lance, capelin, and herring. Larvae feed off a yolk sac for a month; then, as *parr* (actively feeding, freshwater young), eat worms, insects, and other small animals.

PARASITES AND DISEASE: Protozoans, trematodes (numerous spp.), cestodes, nematodes, acanthocephalans, leeches, mollusks, and crustaceans are parasites. Furunculosis, *Aeromonas salmonicida;* vibriosis in culture. Hitra disease (hemorrhagic syndrome) may be caused by more than one agent.

PREDATORS AND COMPETITORS: Eaten by birds and fish as freshwater young, and by large fish and mammals (seals) as saltwater adults.

AQUACULTURAL POTENTIAL: This species is farmed on both coasts of Canada, though toxic algae have caused serious mortalities on Canada's Pacific Coast, at Vancouver. Atlantic salmon farms have also been operated on both U.S. coasts, Maine and New Hampshire in the east and Washington in the west, and raceway culture of early stages of Atlantic salmon is conducted in some inland areas. Both countries experience conflict between commercial salmon fishers and salmon farmers. The state of Alaska is opposed to farming Atlantic salmon in Alaskan waters, in part to prevent losses to fisheries. Very successful Atlantic salmon net-pen culture is reported from Norway, Scotland, Iceland, and the Faroe Islands. Unfortunately, increased worldwide production has developed a market crisis, causing a sharp drop in the per-pound price of salmon.

REGIONS WHERE FARMED AND/OR RESEARCHED: United States and Canada; Australia, British Columbia, Chile, Scotland, Norway, and Sweden.

REFERENCES

Laird, L. M., and T. Needham, eds. 1988. *Salmon and trout farming*. New York: John Wiley.
Sedgwick, S. D. 1988. *Salmon farming handbook*. Farnham (U.K.): Fishing News Books.

SCIENTIFIC AND COMMON NAMES: ***Oncorhynchus tshawytscha*, Chinook salmon**
Other Pacific salmon: Coho salmon (*O. kisutch*), pink salmon (*O. gorbuscha*), and steelhead trout (recently reclassified as a salmon, *O. mykiss;* see entry under "Trout")

DESCRIPTION AND DISTINCTIVE CHARACTERISTICS: Large, powerful, anadromous fish. The body is elongate, head conical, and mouth terminal; there is an adipose fin.

HABITAT RANGE: California to Alaska in freshwater rivers and streams; marine life stage in the north Pacific Ocean is generally east of the International Date Line.

REPRODUCTION AND DEVELOPMENT: Females build nests in stream gravel, generally laying 3,000–5,000 eggs, which are fertilized externally by the male; the nest is then covered over with gravel. Young spend anywhere from a week to a year in fresh water as fry and fingerlings before going to sea, later to return to the freshwater stream whence they originated. Generally, Pacific salmon die after spawning (see the section "Salmonids" on p. 117).

AGE AND GROWTH: After spending up to a year in fresh water, chinook salmon may remain at sea for three or four years. They may reach over 4 ft in length and weigh 90 lb.

FOOD AND FEEDING: As young in freshwater, chinook salmon eat plankton; as adults at sea they eat herring, small shrimp, needlefish, and pilchards.

PARASITES AND DISEASE: There are about 30 known species of parasites, including protozoans, trematodes, cestodes, nematodes, acanthocephalans, and copepods. Furunculosis, columnaris disease, enteric redmouth, and vibriosis are caused by bacteria; coldwater disease is due to a protozoan. Infectious hematopoietic necrosis virus (IHNV) is a serious pathogen of chinook salmon in the Pacific Northwest.

PREDATORS AND COMPETITORS: Species are cannibalistic when young. Adults are eaten by fur seals and large finfish.

AQUACULTURAL POTENTIAL: Salmon in general have a high potential for successful aquaculture because of substantial background on biological aspects, spawning, and so on. They are easy to culture and hold in captivity, and markets are strong.

REGIONS WHERE FARMED AND/OR RESEARCHED: Washington, Oregon; Canada; Japan and Chile.

REFERENCES

Laird, L. M., and T. Needham, eds. 1988. *Salmon and trout farming*. New York: John Wiley.
Sedgwick, S. D. 1988. *Salmon farming handbook*. Farnham (U.K.): Fishing News Books.

Seatrout

SCIENTIFIC AND COMMON NAMES: ***Cynoscion nebulosus,*** **Spotted seatrout**
Also called spotted weakfish, speckled trout.

DESCRIPTION AND DISTINCTIVE CHARACTERISTICS: A rather large fish (reaching about 28 in.) related to the croakers and drums (Family: Sciaenidae). It is dark gray above with bluish reflections and numerous round, black spots irregularly scattered on the back and upper sides.

HABITAT RANGE: From Cape Cod southward around the Gulf of Mexico to the Gulf of Campeche, Mexico. Optimum temperature range is 59–81 °F. Seatrout are tolerant to low salinities (520 ppt), and frequent estuarine and brackish waters, as well as shallow-water grass flats. Migration is very limited.

REPRODUCTION AND DEVELOPMENT: Females become mature at about 9.5 in., males at about 9 in. Spawning takes place in deep channels during late spring and throughout the summer, lasting in some areas until November. Large females may release over 1.5 million eggs during a spawning season.

AGE AND GROWTH: Females are believed to live at least seven years and males about six.

FOOD AND FEEDING: Are carnivorous, eating mostly shrimp and small fish. Small adult seatrout tend to eat proportionately more shrimp than fish, whereas large adults eat more fish than shrimp. Algae and marine plants, mollusks, and crabs constitute a small portion of the stomach contents.

PARASITES AND DISEASE: Dinoflagellates, protozoans, cestodes, monogeneans, nematodes, annelids, copepods, and isopods (numerous species in each group).

PREDATORS AND COMPETITORS: Sharks and large fish are predators.

AQUACULTURAL POTENTIAL: A desirable fish from a marketing standpoint, but no more than research has been done on rearing early stages. The tiny size of the eggs suggests that mass-rearing might prove a stumbling block in a commercial seatrout farming operation.

REGIONS WHERE FARMED AND/OR RESEARCHED: Alabama, Florida, Mississippi, South Carolina, and Texas.

REFERENCES

Overstreet, R. M. 1983. Aspects of the biology of the spotted seatrout, *Cynoscion nebulosus* in Mississippi. *Gulf Research Reports, Supplement* 1:1–43.

Rutherford, E., E. Thue, and D. Buker. 1982. *Population characteristics, food habits and spawning activity of spotted seatrout,* Cynoscion nebulosus, *in Everglades National Park, Florida.* South Florida Research Center Report T-668. Homestead, FL: National Park Service, Everglades National Park.

Shad

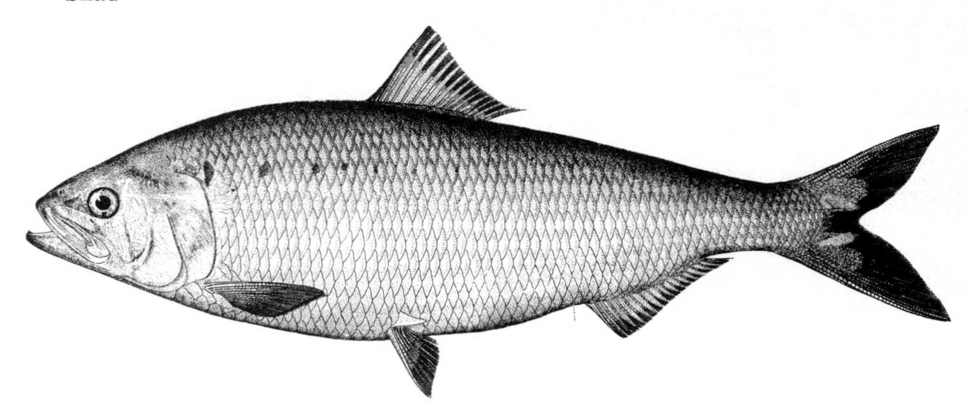

COMMON AND SCIENTIFIC NAME: ***Alosa sapidissima,* American shad** Also called roe shad, white shad, elft.

DESCRIPTION AND DISTINCTIVE CHARACTERISTICS: Typical herringlike fish, with a compressed, *fusiform* (tapered-ends) shape and a deeply forked tail. There is a strongly serrated midline of the belly and weak fins. It is silvery in color, with a bluish-green metallic luster on dorsal side. Its large scales are easily shed.

HABITAT RANGE: Found along the North American Atlantic Coast from St. Lawrence, Canada, to St. John's River, Florida, though most abundant from North Carolina to Connecticut. They have also been successfully transplanted to the Pacific Coast (from the Mexican border to Cook Inlet, Alaska).

REPRODUCTION AND DEVELOPMENT: American shad are anadromous; like salmon, they spend most of their lives in the ocean but return to their home streams to spawn. They spend the first summer of their lives in the streams where they were hatched; at a length of 3–6 in. they move out to sea and remain there until they mature, usually at age 3–4 years. They ascend rivers to spawn as early as November at the southern end of their range in Florida, and as late as July at the northern extreme in Canada. Stream mortality is high at the southern end of their range and lowest at the northern extreme.

AGE AND GROWTH: American shad may reach 14 lb and a length of 30 in. They live about five or six years.

FOOD AND FEEDING: Young eat insects and crustaceans in fresh water; once at sea, they shed their teeth and become plankton feeders using their gill rakers.

PARASITES AND DISEASE: The American shad is parasitized by only a few different parasite species, serving as a host for nematodes, trematodes, acanthocephalans, and copepods.

PREDATORS AND COMPETITORS: Eels and several species of freshwater fish (e.g., carp, catfish, and largemouth bass) feed on juvenile shad. In California, striped bass stomachs contained shad; in North Carolina sea-run shad are eaten by sharks.

AQUACULTURAL POTENTIAL: Shad hatcheries along the Atlantic Coast have been in existence since the late 1800s. Despite large-scale hatching and stocking of young shad in streams during 1880–1950, the populations of shad declined. Early attempts at management of the shad fisheries were typical of the research trend of the time: Depleted wild populations encouraged increased output of hatchery-reared individuals; other factors adversely affecting stock size were left unconsidered. (See the section "Early public aquaculture" in Chapter 2.) Stocks were harmed by pollution, overfishing, and the damming of streams, which prevented their migration to suitable spawning areas. Even releases of millions of hatchery-reared individuals year after year will not halt declines in stock size if they comprise the only management strategy used; but eventually stream pollution was reduced, fish-passage facilities were installed, and fishing was regulated. In addition, more effective hatchery-rearing techniques were developed. Only recently did runs start to increase in some rivers: In the Susquehanna River, record increases in shad abundance were seen over a four-year period (1988–91).

REFERENCES

Mansueti, R., and H. Kolb. 1953. *A historical review of the shad fisheries of North America.* Chesapeake Biological Laboratory Publication no. 97. Solomons, MD: Chesapeake Biological Laboratory.

Walburg, C. H., and P. R. Nichols. 1967. *Biology and management of the American shad and status of the fisheries, Atlantic coast of the United States, 1960.* Special Scientific Report—Fisheries no. 550. Washington, DC: U.S. Fish and Wildlife Service.

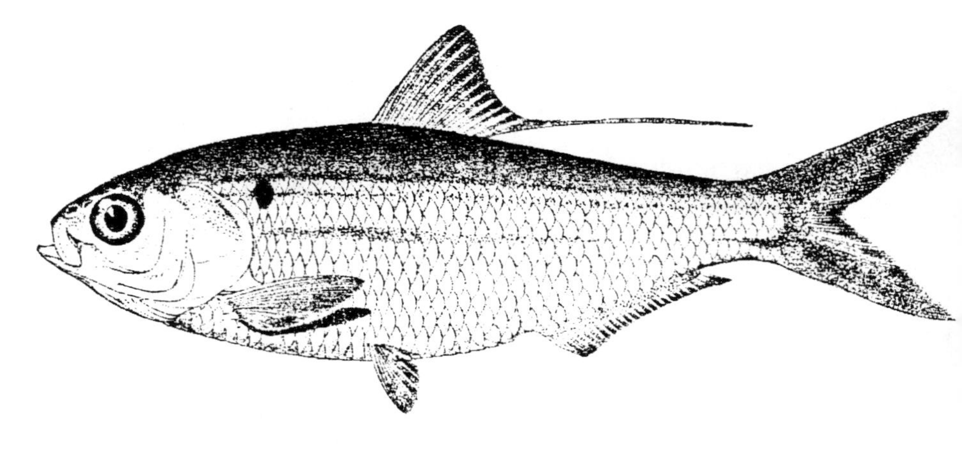

COMMON AND SCIENTIFIC NAME: ***Dorosoma petenense*, Threadfin shad**

DESCRIPTION AND DISTINCTIVE CHARACTERISTICS: The common name of this fish derives from its most conspicuous characteristic: the long dorsal filament that distinguishes it from the American shad. The body is compressed and covered with relatively large scales; the sides are silvery, and the back a bluish-black or olive drab color. There is a dark black oval or round spot behind the upper part of the opercle. The mouth is terminal.

HABITAT RANGE: Threadfin shad range from the Ohio River in Kentucky and southern Indiana westward and southward to Oklahoma, Texas, and Florida; they continue along the Gulf of Mexico to Guatemala and British Honduras. These shad have been transplanted widely in California and in several reservoirs in Hawaii, and are reported from large rivers, bays, reservoirs, lakes, and estuaries over a wide range of salinity (~3–26 ppt). These pelagic fish may form large schools.

REPRODUCTION AND DEVELOPMENT: Spawning begins at less than a year old, and apparently peaks in the spring and fall. Adhesive eggs are released over plants and other submerged objects when the temperature is about 70 °F.

AGE AND GROWTH: Threadfin shad live only about two years. They generally do not reach a length greater than 6 in. in the northern part of their range, but may reach about 8 in. to the south.

FOOD AND FEEDING: Consume unicellular green algae and diatoms; may also eat fry of some fish. They accept an artificial diet readily.

PARASITES AND DISEASE: A bacterium, *Aeromonas liquefaciens,* has been associated with mass mortalities of stressed threadfin shad. A protozoan, *Cryptobia* sp., infects these fish.

PREDATORS AND COMPETITORS: Eaten by a variety of predatory fishes. They have high value as forage fish in the wild, making up as much as 90 percent of the diet of some predatory fishes.

AQUACULTURAL POTENTIAL: Have been used to establish predator–prey relations. The threadfin shad serves as a good food for young sportfish such as largemouth bass, catfish, and crappie. They have rather high fecundity, feed low on the food chain, and grow rapidly.

REGIONS WHERE FARMED AND/OR RESEARCHED: Nebraska Game and Parks Commission, and Texas.

REFERENCES

Ellison, D. G., J. A. Gleim, and D. Kapke. 1983. Overwintering threadfin shad through intensive culture. *Progressive Fish-Culturist* 45(2):90–3.

Texas Aquaculture Association. 1990. *Inland aquaculture handbook.* Austin, TX: Texas Aquaculture Association.

Snapper

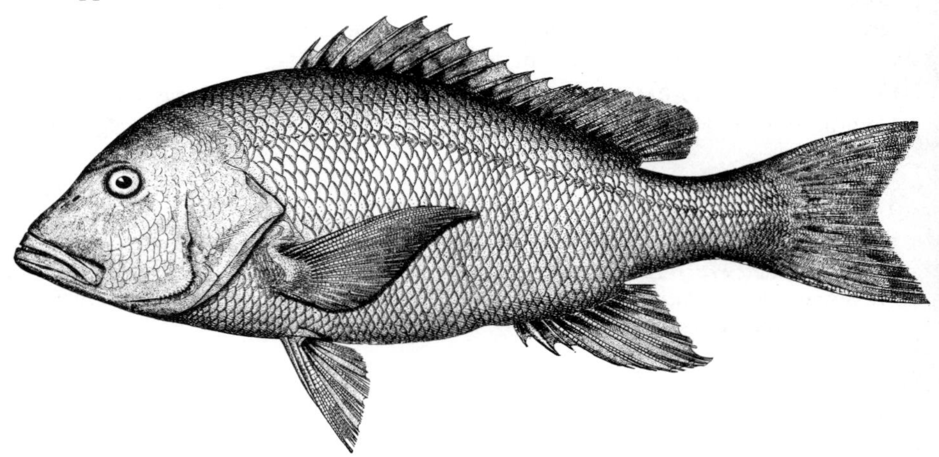

SCIENTIFIC AND COMMON NAMES: ***Lutjanus campechanus,* Red snapper**

DESCRIPTION AND DISTINCTIVE CHARACTERISTICS: A deep-bodied fish with a strong, forked caudal fin. The dorsal fin has about 10 sturdy spines and 14 soft rays, and there is an anal fin with three spines. Color is brick red to scarlet on back and sides.

HABITAT RANGE: In rocky bottoms on continental shelves along the Atlantic Coast north to Cape Hatteras, North Carolina, and in the Gulf of Mexico. May rarely be seen as far north as Massachusetts.

REPRODUCTION AND DEVELOPMENT: Some red snappers become sexually mature at about age 1 year, reaching full maturity at about 2. Young females may release about 200,000 eggs, old females about 19.3 million. Spawning occurs during the summer months, as a rule. Snappers appear to be *multiple spawners,* spawning several times a season.

AGE AND GROWTH: May live to 13 years. Growth is very rapid as juveniles, but slows considerably once full grown. In Louisiana, 13-year-old fish may reach a length of 3 ft and weigh about 26 lb.

FOOD AND FEEDING: Opportunistic polyphagous feeders, eating mollusks, arthropods, and other fishes (eels, Clupeidae, Serranidae, and Syngnathidae have been found in stomachs).

PARASITES AND DISEASE: Leeches attach to the gills.

PREDATORS AND COMPETITORS: Preyed on by sharks.

AQUACULTURAL POTENTIAL: Red snapper is a very popular commercial fish with high demand. Unfortunately, considerable biological research and rearing trials must be made before the aquacultural potential of this species can be determined.

REGIONS WHERE FARMED AND/OR RESEARCHED: Florida, Alabama, Texas; Cuba and Venezuela; Grand Cayman, Cayman Islands; Martinique, French West Indies (*L. analis, L. griseus, L. synagris*).

REFERENCE

Moran, D. 1988. *Species profiles: Life histories and environmental requirements of coastal fishes and invertebrates (Gulf of Mexico). Red Snapper.* Biological Report 82(11.83) TR EL-82-4. Washington, DC: U.S. Fish and Wildlife Service.

Snook

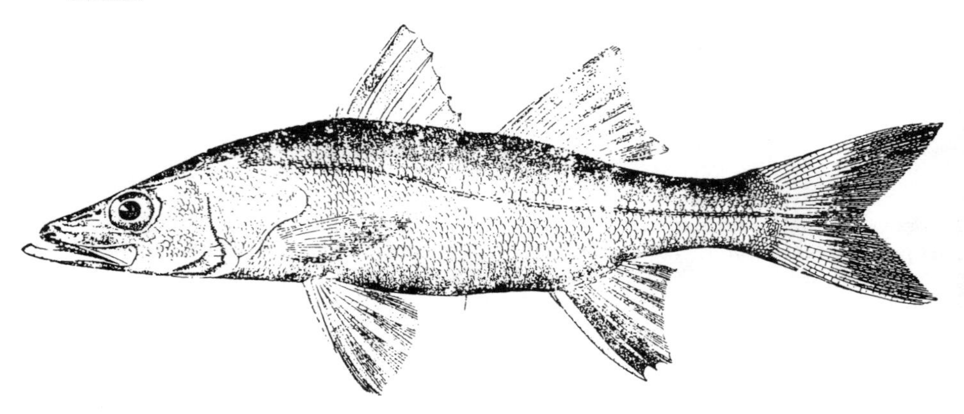

SCIENTIFIC AND COMMON NAMES: ***Centropomus undecimalis,* Snook** Also called the common snook.

DESCRIPTION AND DISTINCTIVE CHARACTERISTICS: A rather large fish, typically with a long concave head, an undershot lower jaw, and two dorsal fins. Sides are silvery with a pronounced black lateral line extending onto the caudal fin.

HABITAT RANGE: Subtropical distribution from North Carolina to Brazil, including the West Indies; in the United States from Texas to Florida. Snook are relatively nonmigratory, require warm water (78–86 °F), and are usually found in estuarine areas.

REPRODUCTION AND DEVELOPMENT: Almost 50 percent of snook studied in Florida reached maturity at 16 in.; nearly all were mature by about 20 in. Spawning takes place in shallow water off sandy beaches in the mouths of saline open-water passes. Fecundity may be as high as 1.4–1.6 million eggs in these summer spawners. There is evidence suggesting that snook, like snapper, may be multiple spawners.

AGE AND GROWTH: May live to age 7 years or more, reaching about 50 lb and a length of 4 ft.

FOOD AND FEEDING: Carnivorous, feeding mainly on fish (largely on mullet) and crustaceans; occasionally cannibalistic.

PARASITES AND DISEASE: An unidentified nematode in the mesenteries and stomach wall may reach a high rate of infection. The nematode *Philometra centropomi* infects the nasal mucosa; in Mexican waters, the trematode *Prosthenhystera obesa* may affect the gall bladder. A virus has also been found in this fish.

PREDATORS AND COMPETITORS: Snook are probably eaten by a variety of wading birds and larger fish that inhabit nursery grounds.

AQUACULTURAL POTENTIAL: Snook is a fine-tasting fish that is very popular with recreational fishers. Hatchery techniques are well established. In Florida considerable effort has been devoted to mass-rearing snook to augment the seriously declining wild populations; whether massive hatchery releases can turn this decline around is still uncertain. This problem becomes especially difficult when the cost–benefit ratio of hatchery construction and operation are considered. It is doubtful that private commercial aquaculture is possible in the United States until more economic information is available. Aquacultural trials in Brazil and Venezuela have been encouraging.

REGIONS WHERE FARMED AND/OR RESEARCHED: State of Florida, Department of Natural Resources; Florida Game and Fresh Water Fish Commission; and the University of Miami, RSMAS, for whom sportfishing groups, individual sportfishers, and miscellaneous groups raised money—and local government agencies provided seed money—for research on snook-rearing methods.

REFERENCES

Marshall, A. R. 1958. *A survey of the snook fishery of Florida, with studies of the biology of the principal species,* Centropomus undecimalis *(Bloch).* Florida State Board of Conservation Technical Series 22. Miami, FL: Marine Laboratory, University of Miami.

Seaman, W., Jr., and M. Collins. 1983. *Species profiles: Life histories and environmental requirements of coastal fishes and invertebrates (South Florida)—Snook.* U.S. Fish and Wildlife Service FWS/OBS-82/11.16. U.S. Army Corps of Engineers TR EL-82-4. Slidell, LA: National Coastal Ecosystems Team, U.S. Fish and Wildlife Service.

Shafland, P. L., and D. H. Koehl. 1980. Laboratory rearing of the common snook. *Proceedings of the Annual Conference Southeastern Association of Fish and Wildlife Agencies* 33:425–31.

Thue, E. B., E. S. Rutherford, and D. G. Buker. 1982. *Age, growth and mortality of the common snook,* Centropomus undecimalis *(Bloch), in Everglades National Park, Florida.* South Florida Research Center Report T-683. Homestead, FL: National Park Service, Everglades National Park.

Tucker, J. W., Jr. 1987. Snook and tarpon snook culture and preliminary evaluation of commercial farming. *Progressive Fish-Culturist* 49:49–57.

Volpe, A. V. 1959. *Aspects of the biology of the common snook,* Centropomus undecimalis *(Bloch) of southwest Florida.* Florida State Board of Conservation Technical Series 31. Miami, FL: Marine Laboratory, University of Miami.

Sturgeon

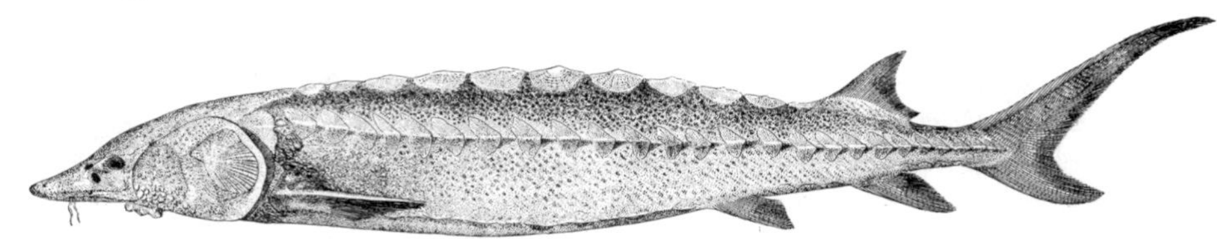

SCIENTIFIC AND COMMON NAMES: ***Acipenser transmontanus*, White sturgeon**
Related: Atlantic sturgeon, *Acipinser oxyrhynchus.*

DESCRIPTION AND DISTINCTIVE CHARACTERISTICS: The body of the white sturgeon is elongate and fusiform, scaleless and whitish; the back is olive to brown. There are five rows of raised *scutes* (external horny or bony plates): two lateral, two ventral, and one dorsal; these are sharp and strong but become dulled in adults. The snout is flattened and blunt when viewed from above. Four barbels are located in advance of the toothless mouth. The dorsal lobe of the tail is curved upward and resembles the tail of a shark. The skeleton is cartilaginous.

HABITAT RANGE: From Mexico to the Aleutian Islands of Alaska. Landlocked, semianadromous, and anadromous populations of white sturgeon exist.

REPRODUCTION AND DEVELOPMENT: Age at first maturity is 9–22 years for males and 11–34 years for females. During each breeding season females may lay between 100,000 and 4.7 million eggs, depending on their size.

AGE AND GROWTH: Growth is slow. These fish may remain in the ocean for as long as twenty years before migrating up rivers to spawn. About three years are required for sturgeon to reach food market size (~12 lb).

FOOD AND FEEDING: The white sturgeon is a bottom feeder that relies on barbels located under the snout for sensing the presence of food items (e.g., insect larvae, worms, clams, snails, and small fish), which it sucks into its toothless mouth. Older, larger fish may eat more fish than younger, smaller ones.

PARASITES AND DISEASE: Eggs parasitized by *Polypodium,* a coelenterate; other coelenterates parasitize the gills, mesenteries, and blood vessels. Nematodes, acanthocephalans, hirudineans, arthropods, and lampreys are also parasites on this species. A viral infection that caused high mortality was reported in one facility.

PREDATORS AND COMPETITORS: White sturgeon compete for food with other benthic feeders. When small they are eaten by a variety of piscivorous fishes.

AQUACULTURAL POTENTIAL: A related species, the Atlantic sturgeon (*Acipinser oxyrhynchus*), was artificially spawned in North America as early as 1875. Until 1902 there were two major problems associated with spawning sturgeons: One was obtaining a supply of simultaneously ripe males and females (since the advanced age of sexual maturity requires that eggs be taken from wild populations); the other was preventing the destruction of eggs by fungal infection. The first successful reproduction of the white sturgeon in captivity took place in 1980. Injections of pituitary glands of carp and sturgeon were used to induce ovulation, and large numbers of young were reared to fingerling size. Improvements in rearing technology are still needed, with considerable research directed to diet improvement for reared sturgeons. The meat of the sturgeon is very high in food value and is often smoked. Quality caviar (sturgeon eggs) commands extremely high prices.

REGIONS WHERE FARMED AND/OR RESEARCHED: Charleston, South Carolina; California (45 farms). Other species of sturgeon are being researched or considered for farming in the United States and Canada, including the Atlantic sturgeon, *A. oxyrhynchus*.

REFERENCES

Anderson, E. R. 1984. *Artificial propagation of lake sturgeon,* Acipenser fulvescens *(Rafinesque), under hatchery conditions in Michigan.* Fisheries Research Report no. 1989. Lansing, MI: Michigan Dept. of Natural Resources, Fisheries Division.

Conte, F. S., S. I. Doroshov, and P. B. Lutes. 1988. *Hatchery manual for the white sturgeon,* Acipenser transmontanus *Richardson, with application to other North American Acipenseridae.* Cooperative Extension University of California, Division of Agriculture and Natural Resources Publication 3322. Oakland, CA: Division of Agriculture and Natural Resources, University of California.

Dadswell, M. J., B. D. Taubert, T. S. Squiers, D. Marchette, and J. Buckley. 1984. *Synopsis of biological data on shortnose sturgeon,* Acipenser brevirostrum *LeSueur 1818.* NOAA Technical Report NMFS 14. Seattle, WA: National Marine Fisheries Service.

Doroshov, S. I. 1985. Biology and culture of sturgeon, Acipenseriformes. In *Recent advances in aquaculture,* vol. 2, ed. J. F. Muir and F. J. Roberts, pp. 251–75. Boulder, CO: Westview Press.

Tilapia

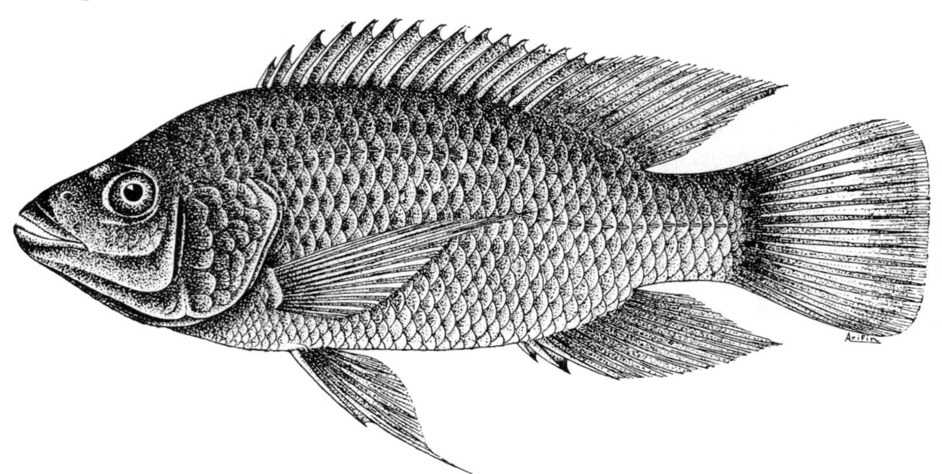

SCIENTIFIC AND COMMON NAMES: ***Tilapia mossambica*, Mozambique tilapia**
Related: *Sarotherodon* spp.

DESCRIPTION AND DISTINCTIVE CHARACTERISTICS: Relatively small, normally drab-colored fish capable of living in fresh or salt water.

HABITAT RANGE: Widely transplanted from Africa, where Lake Malawi is a source for a wide variety of species. There is evidence of tilapia occurring in Egypt in 2,500 B.C. They live in freshwater, brackish water, and marine habitats, generally requiring temperatures of 61–86 °F. Tilapia in streams seek those with slow flow.

REPRODUCTION AND DEVELOPMENT: The males of this *Tilapia* species dig depressions (crude nests) in the bottom and entice females to nest; in some other species, eggs are laid directly on the bottom. Nest builders lay 7,000–8,000 eggs. "Mouth brooders" (*Sarotherodon* spp.) lay over 200 eggs every two months and brood them orally: In some species it is only the females that do so, in others the males; in yet others, both sexes care for the eggs. Young are protected from predators by entering the parent's mouth.

AGE AND GROWTH: Tilapia grow rapidly and may reach over 5 in. in about three months; males grow faster than females.

FOOD AND FEEDING: Omnivorous, but seem to prefer plants, especially *Enteromorpha* (brackish water algae).

PARASITES AND DISEASE: Virus (lymphocystis), protozoans, digenetic trematodes, monogenetic trematodes, cestodes, nematodes, copepods, and isopods.

PREDATORS AND COMPETITORS: Predatory birds and fishes eat tilapia.

AQUACULTURAL POTENTIAL: The potential for this species is not too high. Although their rapid growth and food habits make them a desirable species, they tend to overreproduce and are sensitive to cold temperatures. (Water temperatures are too cold for tilapia in most of the continental United States.) U.S. production has been increasing at about 5 percent annually. Tilapia can be used in polyculture farms in association with milkfish, mullet, catfish, and shrimp. The U.S. market for this fish is weak due to its small size and undesirably drab colors; however, colorful strains (blond, orange-yellow, pink, red, gold, and bronze) have been developed, and a market for tilapia as ornamental fish now exists in addition to that as food for humans. Hybrids have also been developed to produce largely single-sex (male) offspring, thus preventing overreproduction while providing more rapid growth.

REGIONS WHERE FARMED AND/OR RESEARCHED: Florida, California; the Bahamas, Jamaica; Taiwan, the Philippines (total ≈ 30 countries).

REFERENCES

Balarin, J. D. 1979. *Tilapia: A guide to their biology and culture in Africa.* Stirling (Scotland): University of Stirling.

Goldstein, R. J. 1973. *Cichlids of the world.* Neptune City, NJ: T. F. H. Publications.

Pullin, R. S. V., and R. H. Lowe-McConnell, eds. 1980. *The biology and culture of tilapias.* ICLARM Conference Proceedings 7. Manila (Philippines): International Center for Living Aquatic Resources Management.

Smith, I. R., E. B. Torres, and E. O. Tan. 1983. *Philippine tilapia economics.* ICLARM Conference Proceedings 12. Manila (Philippines): International Center for Living Aquatic Resources Management.

Trewavas, E. 1983. *Tilapiine fishes of the genera Sarotherodon, Oreochromis, and Danakilia.* Publication of the British Museum (Natural History) no. 878. London: British Museum (Natural History).

Trout

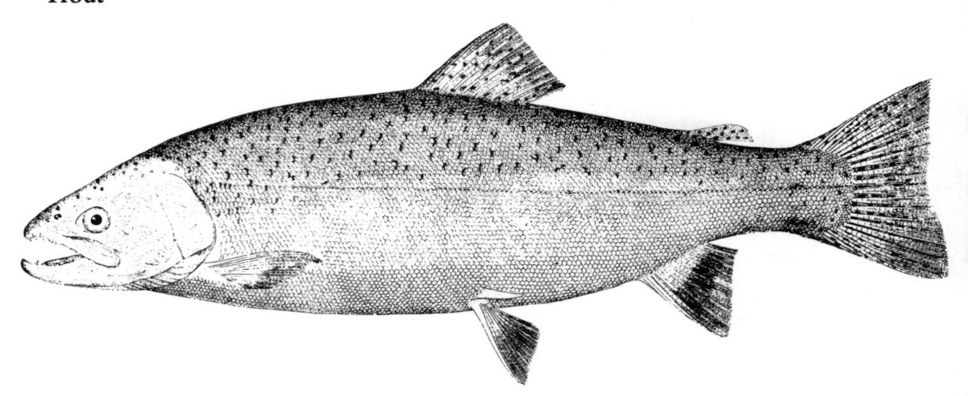

SCIENTIFIC AND COMMON NAMES: ***Oncorhynchus mykiss*****, Steelhead trout**

This salmonid was formerly classified as *Salmo gairdneri* and called the "rainbow trout." Members of the species that spend their entire life in freshwater streams are still commonly called rainbow trout; steelhead trout are those that spend most of their life in fresh water but migrate to the ocean during their maturation period.

DESCRIPTION AND DISTINCTIVE CHARACTERISTICS: This back is olive colored, and there is a red band on either side of the body.

HABITAT RANGE: The colder waters of North and South America, generally in oxygen-rich, fast-flowing streams. Optimum range is 50–65 °F; growth is substantially reduced at 32°–50°, and 86° is lethal. Transplanted over a wide range from streams in northern California and later from Washington and Oregon.

REPRODUCTION AND DEVELOPMENT: Females produce 2,000–3,000 eggs during the spawning season. Ripe females can be spawned artificially, and fry survival rates are high. Early life is spent as yolk-sac larva; feeding begins as soon as the nourishment in the sac is consumed.

AGE AND GROWTH: Can be raised to about 1 lb or so (market size) in about 10–12 months.

FOOD AND FEEDING: Eat insects, flies, crustaceans, worms, small fish, and salmon eggs. Pelleted foods available usually contain wheat, fish meal, whey, cottonseed meal, and yeast.

PARASITES AND DISEASE: Fungi, protozoans, trematodes, cestodes, nematodes, acanthocephalans, leeches, mollusks, gordiaceans (which resemble nematodes), and crustaceans are all parasites of this species. Infectious hematopoietic necrosis virus (IHNV) is a serious trout pathogen. A protozoan causes whirling disease, which results in high mortality in culture.

PREDATORS AND COMPETITORS: Birds, snakes, turtles, rodents, raccoons, muskrats, and other fish eat steelhead trout.

AQUACULTURAL POTENTIAL: Trout farming started in the United States during the 1860s; since its beginnings, the principal species farmed has always been the rainbow trout (now called steelhead trout unless raised entirely in fresh water). The market for trout was weak until the 1950s, and improved considerably during the mid- to late 1960s. By 1970 good-quality dry feeds were available. Trout farming is well established in many states due to the abundant literature on trout biology and the long history of farming trials. Research critical to successful farming included studies on artificial selection, disease identification, prevention, and control, and the production of dry pelleted feeds. (See also the section on U.S. freshwater trout farming in Chapter 2.) Steelhead trout require high-quality, oxygen-rich, rapidly flowing water, as do most salmonid fishes.

REGIONS WHERE FARMED AND/OR RESEARCHED: Northern United States, South America, and Scandinavia.

REFERENCES

Laird, L. M., and T. Needham, eds. 1988. *Salmon and trout farming*. Chichester (U.K.):Ellis Horwood.

Sedgwick, S. D. 1985. *Trout farming handbook*, 4th ed. Farnham (U.K.): Fishing News Books.

Stevenson, J. P. 1987. *Trout farming manual*, 2nd ed. Farnham (U.K.): Fishing News Books.

Tuna

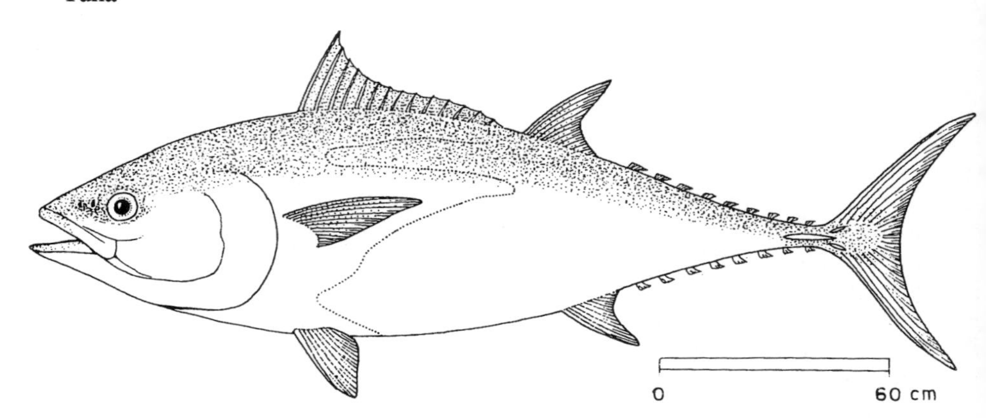

SCIENTIFIC AND COMMON NAMES: ***Thunnus thynnus,* Bluefin tuna** Also called Northern bluefin tuna.

DESCRIPTION AND DISTINCTIVE CHARACTERISTICS: A large, powerful pelagic fish whose streamlined body has a large, forked caudal fin, just in advance of which are nine or so dorsal and anal finlets. The large eye and mouth confirms the predatory nature of this rapid swimmer (40–50 mph). The lower sides of the body and underside are silvery white; the dorsal side is deep blue or green.

HABITAT RANGE: Epipelagic (0–600-ft depth), occasionally coming near shore, though the majority of its life is spent roaming widely over the open sea. In the Atlantic the bluefin is distributed from Labrador and Newfoundland south into the Gulf of Mexico and the Caribbean Sea, and on the western Atlantic to Brazil and Venezuela. Their range extends further north on the eastern Atlantic up to Lofoten Island off Norway, and south to the Canary Islands and the Mediterranean. The Pacific subspecies occurs on the western Pacific in the vicinity of the Japanese side of the Sea of Okhotsk south to the northern Philippines; in the eastern Pacific it extends from the Gulf of Alaska to southern California.

REPRODUCTION AND DEVELOPMENT: Bluefin tuna become mature at age 4–5 years. Spawning occurs in the Gulf of Mexico and the Mediterranean Sea, with as many as 10 million eggs released by a female per season. Bluefin tuna have been spawned in captivity in Japan.

AGE AND GROWTH: At roughly age 3 years bluefin are about 7 ft long. They may attain a weight of over 800 lb.

FOOD AND FEEDING: Eat a variety of small schooled fishes such as anchovies, sauries, hakes, and squids; also may eat some crabs. In captivity, they will take artificial feeds.

PARASITES AND DISEASE: In culture, bacterial and parasitic diseases, together with nutritional diseases, call for better rearing procedures and improved diets.

PREDATORS AND COMPETITORS: Planktonic larvae and small juveniles are preyed upon; in later life, few if any predators are large or fast enough to catch bluefin.

AQUACULTURAL POTENTIAL: Japan has experimented with several species of tuna, including bluefin, yellowfin, dogtooth, bonito, skipjack, kawakawa, frigate, and bullet tuna. They believe that the bluefin tuna has the greatest potential because of its astonishingly high market value and rapid growth rate. Young bluefin have been caught at sea, taken to enclosures (such as closed-off coral atolls and lagoons), and raised there to market size. Floating pens of the sort used for rearing Atlantic salmon have also been tried with bluefin. The very expensive feeds required for their maintenance and growth is a major stumbling block in commercial aquaculture. It has been suggested that large enclosures be fertilized to natural food chains to feed the tuna, but even this is very expensive and may well upset the ecological balance of the local habitat. Rearing trials have been frustrated due to the lack of mature eggs. The tuna market suffered a serious setback when mercury levels were found to be high in this species. This did not harm the Japanese market to the extent it did the American and Canadian markets; however, U.S. demand has rebounded, and tuna consumption is the highest of all seafoods.

REGIONS WHERE FARMED AND/OR RESEARCHED: Nova Scotia and Japan.

REFERENCES

Buchanan, L. 1977. Ranching Atlantic bluefin. *Sea Frontiers* 23(3):172–80.

Harada, T. 1979. Present status of mariculture study in Japan for tuna species. In *Le thon rouge en Mediterranee: biologie, peches et aquaculture,* by F. X. Bard and J. Y. Le Gall, pp. 163–7. Actes de colloques du CNEXO, no. 8. Brest (France): Centre Oceanologique de Bretagne.

Walleye

SCIENTIFIC AND COMMON NAMES: ***Stizostedion vitreum,*** **Walleye** Also called yellow pikeperch, walleyed pike, glasseye.

DESCRIPTION AND DISTINCTIVE CHARACTERISTICS: Two dorsal fins, the first with spines. Eyes are prominent, with a light-gathering layer to aid in capturing prey fish in dim light. Color may be dark silver ranging to dark olive brown mottled with brassy specks.

HABITAT RANGE: Found over sandy bottoms in clear waters, more often in lakes than in streams, walleye range from New York State across the Great Lakes and into the Mississippi Valley to Minnesota and Canada. Introduced into the Tennessee Valley, they are now found as far south as southern Alabama and Georgia. There are even occasional accounts of walleye in brackish waters.

REPRODUCTION AND DEVELOPMENT: Spawning takes place over gravel or rocky bottoms (without building a nest) when water temperatures in streams and lakes are 38–44 °F (April or May). A large female may release over 400,000 eggs.

AGE AND GROWTH: May reach over 40 in. in length and weigh 25 lb. Under artificial conditions, walleye reach maturity as early as age 2 years.

FOOD AND FEEDING: Eat minnows, bass, and yellow perch; walleye larvae eat zooplankton. Feeding occurs in turbid lakes during the day and in clear waters at sunset and sunrise.

PARASITES AND DISEASE: Parasites include protozoans, trematodes, cestodes, nematodes, acanthocephalans, leeches, mollusks, and crustaceans.

PREDATORS AND COMPETITORS: Fry are preyed on by a number of invertebrate species and larger fish. Birds prey on walleye in grow-out ponds. If sufficient food is not available, walleye become cannibalistic.

AQUACULTURAL POTENTIAL: Walleye is a very popular game and commercial fish throughout its range. The demand for fingerlings for stocking natural waters and farm ponds greatly exceeds the supply produced in state, provincial, federal, and private culture facilities. As with many other freshwater fish in North America, rearing techniques were developed well before the turn of the century and are essentially the same today. Using wild-caught adults, it is easy to strip the eggs and sperm and fertilize them artificially. Unlike some other freshwater species, mass-rearing of fry at the time they begin feeding is uneconomical under intensive culture conditions. The standard procedure for walleye is to raise them from fry to about a length of 1 in. in earthen ponds fertilized to produce heavy plankton growth. They can then be placed in troughs, tanks, or floating pens for grow-out. They eat pelleted feed.

REGIONS WHERE FARMED AND/OR RESEARCHED: Minnesota Department of Conservation; Michigan Department of Natural Resources; New York, Wisconsin, and Pennsylvania.

REFERENCES

Buttner, J. K. 1989. Culture of fingerling walleye in earthen ponds; state of the art 1989. *Aquaculture Magazine* 15(2):37–46.

Ebbers, M. A., P. J. Colby, and C. A. Lewis. 1988. *Walleye–sauger bibliography*. St. Paul, MN: Minnesota Department of Natural Resources.

Stickney, R. R., ed. 1986. *Culture of nonsalmonid freshwater fishes*. Boca Raton, FL: CRC Press.

Yellowtail

SCIENTIFIC AND COMMON NAMES: ***Seriola quinqueradiata,* Yellowtail**

DESCRIPTION AND DISTINCTIVE CHARACTERISTICS: The yellowtail is in the family Carangidae (amberjacks) and is related to the pompano and jacks. It is a pelagic fish and able to swim at high speeds.

HABITAT RANGE: Pacific Ocean, from the southern part of Kamchatka Peninsula to the Province of Siberia, in waters of about 62–84 °F.

REPRODUCTION AND DEVELOPMENT: There is a large spawning ground in the East China Sea. Larval fish attach themselves to drifting seaweed until they reach a length of about 4 in., moving inshore on currents carrying the seaweed. There they remain until they are age 2 years and 16 in. in length, when they begin to migrate over great distances in the open sea.

AGE AND GROWTH: Small juveniles from the wild reared in cages will reach market size (2–3 lb) in a year. During their second year they will reach 4–6 lb. Maximum size of wild yellowtail is about 18 lb.

FOOD AND FEEDING: In the wild, yellowtail feed on mackerel, horse mackerel, sardines, and squids. In confinement, they are fed minced sand lances, anchovies, and noncommercial species of shrimp. The larvae eat plankton until they reach about 4 in. in length.

PARASITES AND DISEASE: Fry may develop calcium deposits in their urinary bladder ("bladder stones"); an ectoparasitic flatworm, *Bendenia seriole,* may harm the skin of the fish; a trematode of the gills, *Axine heterocerca,* may cause fatal anemia; and, as in many fish, vibriosis (bacterial disease) occurs frequently. Microsporidiosis, fungi, copepods, and various nutritional diseases cause stock losses. "Back splitting disease" occurs in some rearing facilities.

PREDATORS AND COMPETITORS: Fish and birds prey on yellowtail. It is important to grade small yellowtail by size as they are very cannibalistic.

AQUACULTURAL POTENTIAL: Culturing of yellowtail in Japan until 1928 was small scale; the large-scale culture today is still dependent on the collection of fry and fingerlings from natural spawning. Although mass-rearing of yellowtail fry has been perfected, fry for grow-out are collected from wild fish. The availability of large schools of small yellowtail close to shore and the large areas in the inland sea for grow-out make this a very successful operation in Japan. Several attributes of the yellowtail make them a desirable fish for culture: They are easy to feed, fast growing, and have a high market value. Also, despite their wide-ranging behavior and rapid swimming in nature, they can be confined in land enclosures, net pens, or cages with ease. About 50 percent of the cost of rearing yellowtail is in the feed.

REGIONS WHERE FARMED AND/OR RESEARCHED: In Japan, where they are raised in netted-off areas of the inland sea and in cages, yellowtail is far and away the most important cultured species. Although this species does not occur in North America, it is included here because of the phenomenal success of a farmed species that depends on wild fry, and because other members of this family (Carangidae) do occur in North America. Some of these species (e.g., pompano) have been raised in aquacultural facilities; others (e.g., permit [*Trachinotus falcatus*] and amberjack [*Seriola* spp.]) are being considered as aquacultural candidates.

REFERENCES

Davy, F. B. 1990. Mariculture in Japan. 1—Development of an industry. *World Aquaculture* 21(4):36–47.

Matsusato, T. 1984. Present status and future potential of yellowtail culture in Japan. In *Proceedings of the Seventh U.S.–Japan Meeting on Aquaculture, Marine Finfish Culture, Tokyo, Japan, 1978*, ed. C. J. Sindermann, pp. 11–16. NOAA Technical Report NMFS 10. Seattle, WA: National Marine Fisheries Service.

Nakada, M., and T. Murai. 1991. Yellowtail aquaculture in Japan. In *CRC Handbook of mariculture, Vol. II: Finfish aquaculture*, ed. J. P. McVey, pp. 55–72. Boca Raton, FL: CRC Press.

Reptiles

Reptiles are cold-blooded vertebrates that are usually covered with scales or bony plates. They are most abundant in warmer regions of the world. Unlike amphibians, reptiles breathe with lungs throughout their entire lives, and their eggs are laid on land, protected by a shell. They may be completely terrestrial or semiaquatic; of course, those raised by aquaculturists spend a portion of their lives in and around aquatic environments. Crocodiles and alligators are raised for their meat and skin (leather); sea turtles for a variety of commercial products, mainly shells and meat; and brackish and freshwater turtles (including terrapins) for pets or for human consumption.

Students learning about reptiles find several confusing terms: "turtle," "tortoise," and "terrapin." *Turtle* is often applied to semiaquatic and marine species (some of which supply a variety of commercial products, mainly shells and meat). *Terrapin* applies to certain hard-shelled species that live in brackish and fresh water and have market value for their meat and as pets. *Tortoises* are strictly terrestrial species.

REFERENCES

Ernst, C. H., and R. W. Barbour. 1989. *Turtles of the world.* Washington, DC: Smithsonian Institution Press.

Rebel, T. P. 1974. *Sea turtles and the turtle industry of the West Indies, Florida, and the Gulf of Mexico,* rev. ed. Coral Gables, FL: University of Miami Press.

Ross, C. A., ed. 1989. *Crocodiles and alligators.* New York: Facts on File.

Webb, G. J. W., S. C. Manolis, and P. J. Whitehead, eds. 1987. *Wildlife management: Crocodiles and alligators.* Chipping Norton, N.S.W. (Australia): Surrey Beatty.

Alligator

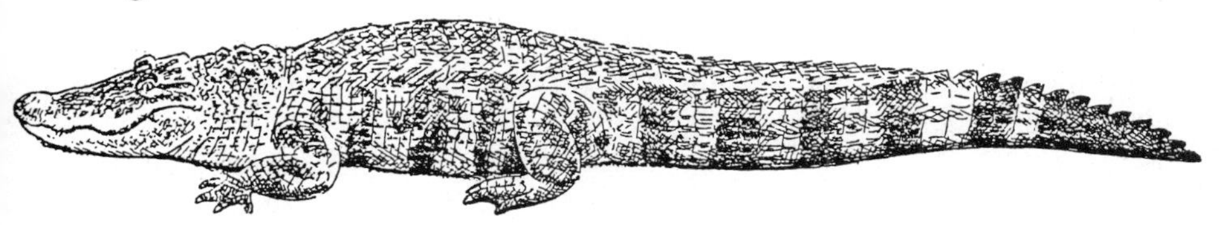

SCIENTIFIC AND COMMON NAMES: ***Alligator mississippiensis*, American alligator**

DESCRIPTION AND DISTINCTIVE CHARACTERISTICS: A large aquatic reptile with a broad head, long body and tail, and dark-colored, heavy skin. The snout is less tapered than the crocodile's, and the undershot jaws have strong, spiked, replaceable teeth.

HABITAT RANGE: Fresh and salt water in tropical and subtropical wetlands.

REPRODUCTION AND DEVELOPMENT: Alligators reach sexual maturity at about age 6. Wild females build nests 3–4 ft high and 6–8 ft wide. Eggs (40–45 per clutch) are laid in late June and generally hatch about 65 days later. At high temperatures (~90 °F) the brood is all males; at lower temperatures, varying numbers of females are also produced.

AGE AND GROWTH: In the wild, alligators grow at a rate of about a foot per year. In farms under ideal conditions, including heated enclosures, they may reach 6 ft (market size) in 2.5 years. Alligators usually reach a length of 13 ft and weigh 600 lb.

FOOD AND FEEDING: Natural food of young consists of insects, small fish, and crustaceans. Adults have been reported to eat anything that moves and many things that do not, including rocks and human litter (soda bottles, shotgun shells, etc.). They can go for extended periods without eating and will gorge themselves when food is plentiful. Artificial feeds have been developed for use by alligator farmers.

PARASITES AND DISEASE: Diseases are produced by abnormal calcium metabolism (metabolic bone disease), which is thought to be caused by being fed boneless meat and receiving inadequate sunlight. Gout, another nutritional disease, causes hind-limb paralysis. Vitamin deficiencies can cause debilitation and occasionally death (by drowning). Shock may result from low blood sugar levels. Two species of digenetic trematodes and a nematode species have been found in the intestine of the American alligator; a third digenetic trematode has been reported in the lungs and liver. A species of parasitic arthropod and some species of parasitic fungi have been reported from alligators in Florida. A disease called "brown spot" causes blemishes and holes on hides; its cause is not known, but it is suspected to be either of viral or bacterial origin. The degree of damage to the hide caused by brown spot may not be known until the hide is tanned.

PREDATORS AND COMPETITORS: Young alligators are preyed upon by snakes, turtles, raccoons, fish, and birds (herons). Large alligators are cannibalistic on the young.

AQUACULTURAL POTENTIAL: Alligators are raised for their fine leather and as a source of meat for human consumption; some facilities also take advantage of their tourist value. The international alligator trade, both legal and illegal, controls the demand and value in the domestic market. Alligator hide prices had reached almost $50/ft, but lessened demand and increased production recently dropped the price to about $17/ft. Infertility of breeding stock, high mortality in the embryonic stages, and diseases (mostly caused by improper diet) reduce production of farmed alligators.

REGIONS WHERE FARMED AND/OR RESEARCHED: Florida (25) farms and Louisiana (120) farms. Expensive alligator research has been carried out by Florida, where alligator farming is considered to be advantageous to the environmental and economic interests of the state.

REFERENCES

Alligators, steamy comfort boosts growth rate. 1988. *Aquaculture Magazine* 14(6):26–9.

Avault, J. W. 1985. The alligator story. *Aquaculture Magazine* 11(4):41–4.

Sheldon, B., and T. Joanen. 1986. Alligators: Louisiana's newest glamour crop. *Louisiana Conservationist* 38(6):4–7.

Webb, G. J. W., S. C. Manolis, and P. J. Whitehead, eds. 1987. *Wildlife management: Crocodiles and alligators.* Chipping Norton, N.S.W. (Australia): Surrey Beatty.

Crocodile

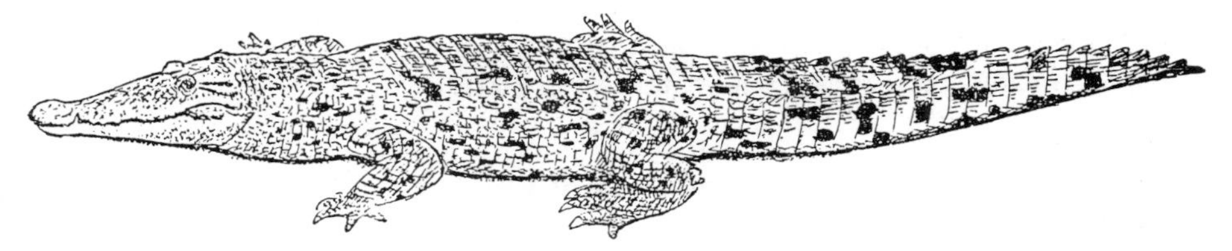

SCIENTIFIC AND COMMON NAMES: ***Crocodylus acutus,* American crocodile**

DESCRIPTION AND DISTINCTIVE CHARACTERISTICS: Like alligators, crocodiles have thick, horny skin made up of scales and plates, and a long, narrow tail, which they use to scull through the water. With their eyes located on top of their head, they are able to float low in the water yet keep a sharp lookout for possible prey. Their teeth are strongly developed, and one enlarged pair in the lower jaw is evident even when their long jaws are closed.

HABITAT RANGE: Saltwater bays, mangrove swamps, and tidal wetlands; salt and fresh waters from southern Florida through numerous Caribbean islands, Mexico, and Colombia.

REPRODUCTION AND DEVELOPMENT: Unlike in some other crocodile species (or alligators), the female does not build a nest; instead, she lays 20–60 hard-shelled eggs in a small excavation made with her snout. Remaining close to the buried eggs, the female digs the hatching young out of the substrate when they emit a chirping sound.

AGE AND GROWTH: Adult length is 16–20 ft.

FOOD AND FEEDING: Young in marine environment feed on crabs, shrimp, insects, and fish. Juveniles and adults eat turtles, birds, snakes, and some mammals.

PARASITES AND DISEASE: Crocodiles of various species from various geographic locations harbor viruses, bacteria, funguses, protozoans, and several species of helminths. Diet deficiencies also cause several syndromes and diseases.

PREDATORS AND COMPETITORS: Young are eaten by raccoons and wild pigs. Some predators are able to locate and eat the buried eggs.

AQUACULTURAL POTENTIAL: Skin is valuable in the manufacture of leather items (wallets, shoes, etc.). Some of the diseases, especially those caused by metabolic deficiencies and lack of preventive methods, may seriously affect the success of crocodile farming. A reliable source of inexpensive and suitable food is an important consideration in crocodile culture. Some species are on the endangered list, and special arrangements must be made to obtain permission to raise them in captivity.

REGIONS WHERE FARMED AND/OR RESEARCHED: Cuba, Australia, Thailand, and Zimbabwe; researched but not farmed in Florida.

REFERENCES

Montague, J. J. 1991. Culture of crocodilians. In *Production of aquatic animals; crustaceans, molluscs, amphibians and reptiles,* ed. C. E. Nash, pp. 209–23. Amsterdam: Elsevier.

Ritchie, T. 1989. Marine crocodiles. *Sea Frontiers* 35(4):212–19.

Webb, G. J. W., S. C. Manolis, and P. J. Whitehead, eds. 1987. *Wildlife management: crocodiles and alligators.* Chipping Norton, N.S.W. (Australia): Surrey Beatty.

Terrapin

SCIENTIFIC AND COMMON NAMES: ***Malaclemys terrapin,* Diamond-back terrapin**
The diamond-back has been divided into seven subspecies; two of these (the Chesapeake and the Carolina) occur along the U.S. Atlantic Coast, and three (the Florida, Louisiana, and Texas terrapins) in the northern Gulf of Mexico.

DESCRIPTION AND DISTINCTIVE CHARACTERISTICS: The upper shell has a shield of concentric, diamond-shaped grooves and ridges; hence the common name.

HABITAT RANGE: Diamond-back terrapin inhabit shallow salt- and brackish-water bays, estuaries, and swamps. They do not normally live in fresh water, but can endure it for long periods of time. The total range of the five species covers the U.S. Atlantic and Gulf coasts from Cape Cod to Texas.

REPRODUCTION AND DEVELOPMENT: Females become sexually mature at age 5 years, and may attain a length of 5.5 in.; males usually do not get larger than 4.75 in. Egg production can vary greatly from season to season, females producing 5–35 eggs per season (about 12 on average). Mating activity is greatest during the spring, egg laying begins in May, and hatching occurs from August through October.

AGE AND GROWTH: Some subspecies may live 25 years or more and reach a shell length of 9 in.

FOOD AND FEEDING: Wild terrapins eat small mollusks and crustaceans. In captivity, they eat cut-up fish, shucked oysters, crabs, and clams.

PARASITES AND DISEASE: Sores caused by bacterial infection and "soft shell" disease both cause mortality in culture facilities. Softening of bones occurs in captive terrapins, as does blindness caused by swelling of the eyelids.

PREDATORS AND COMPETITORS: In the wild, young are eaten by large fish, alligators, rats, and raccoons.

AQUACULTURAL POTENTIAL: Terrapins are valuable as food, with the Chesapeake, Carolina, and Texas terrapin subspecies preferred (in that order) over the others. Flesh is considered a delicacy and has a high market value. *Malaclemys terrapin* is farmed commercially.

REGIONS WHERE FARMED AND/OR RESEARCHED: North Carolina, South Carolina, and Texas.

REFERENCES

Hildebrand, S. F. 1929. Review of experiments on artificial culture of diamond-back terrapin. *Bulletin of the United States Bureau of Fisheries* 45: 25–70.

Hildebrand, S. F., and C. Hatsel. 1926. *Diamond-back terrapin culture at Beaufort, N.C.* U.S. Bureau of Fisheries Economic Circular no. 60. Washington, DC: U.S. Bureau of Fisheries.

SCIENTIFIC AND COMMON NAMES: ***Chelonia mydas*, Green sea turtle**
Related: Hawksbill turtle, *Eretmochelys imbricata.*

DESCRIPTION AND DISTINCTIVE CHARACTERISTICS: This large marine turtle has a single pair of scales on top of its head, between the eyes. The *carapace* (dorsal part of the shell) has a smooth, not serrated, margin and is brown with mottled markings. Limbs are paddle-shaped, and have a single claw on each.

HABITAT RANGE: The green turtle is migratory; its geographic range is confined to tropical and subtropical latitudes.

REPRODUCTION AND DEVELOPMENT: Mating takes place well before egg laying. Eggs are laid in sand on beaches and covered; upon hatching, the young crawl to the sea, where they live until they return to the same beach to spawn.

AGE AND GROWTH: This is a huge animal that may reach 850 lb and may live at least 15 years.

FOOD AND FEEDING: Green sea turtles are mainly herbivores, feeding on a variety of sea grasses, including turtle grass (*Thalassia*). They also eat small mollusks and crustaceans.

PARASITES AND DISEASE: A virus causes gray patch disease, leaving lesions on the skin and shell; bacteria; a protozoan has been found in the intestinal tract (coccidian disease). Bites from other turtles may result in skin ulcers (ulcerate dermatitis). A small crab and amphipods may be commensals.

PREDATORS AND COMPETITORS: Some mammals dig up turtle nests and eat the eggs; young turtles are eaten by sea gulls when they leave the nest.

AQUACULTURAL POTENTIAL: Turtle meat, oil, and shell can be used with little waste; however, a large investment in time and money, the slow growth of sea turtles, and the difficulty of finding a reliable, inexpensive source of young turtles are major disadvantages. Early turtle farming trials in the Caribbean caused

Similar to the green sea turtle is the hawksbill turtle (*Eretmochelys imbricata*).

considerable concern because they were supplied with eggs and young from wild parents collected on beaches of neighboring islands. Furthermore, in recent years the turtle market has suffered a serious setback because green turtles have been declared an endangered species. Although trade is still permitted in farmed or ranched specimens of endangered species, certain regulatory controls must be satisfied. Moreover, there is emotional pressure against trading in species whose wild populations have been declared threatened or endangered by the Convention on International Trade in Endangered Species of Wild Fauna and Flora (CITES).

REGIONS WHERE FARMED AND/OR RESEARCHED: Grand Cayman Islands; Australia.

REFERENCES

Rebel, T. P. 1974. *Sea turtles and the turtle industry of the West Indies, Florida and the Gulf of Mexico*, rev. ed. Coral Gables, FL: University of Miami Press.

Wood, F. 1991. Turtle culture. In *Production of aquatic animals; crustaceans, mollusks, amphibians and reptiles*, ed. C.E. Nash, pp. 225–34. Amsterdam: Elsevier.

4

Nonfood Species

We include in this chapter species of plants and animals from a wide range of phyla that are cultured either as foods for aquacultural species, for bait to catch sportfish, for medical research or environmental assays, as ornamental species, for chemical (carrageen) production, or for miscellaneous market products (e.g., pearls, sponges). As mentioned in the Introduction to this book, some species are raised for more than one market: Summaries of such species are normally given in the discussion of their principal use, to which the reader is directed for additional information.

FOODS FOR AQUACULTURAL SPECIES

The different groups of animals raised in aquacultural facilities—mollusks, crustaceans, and finfish—consume many different foods. Even within these groups there is considerable variation in feeding: Some species are lifelong plankton feeders, whereas others consume plankton only during their early life. As adults, some crustaceans and finfish eat plants only (herbivores), others flesh only (carnivores), and still others eat both plants and flesh (omnivores).

The list of plants and animals that are or may be used as fodder in marine and freshwater aquaculture is lengthy. A good discussion of cultured fish foods is given in McLarney's *The freshwater aquaculture book* (1984), which presents a table of animals usable as foods in aquaculture, broken down by size of food organisms, and distinguishes between those presently being used and those with potential; the latter require new or modified technology and mass-rearing trials before they are usable for aquaculture. As examples of large animals presently cultured for food, he lists earthworms, mealworms, and minnows; large animals that might be suitable include cockroaches, grasshoppers, and crickets. In the midsized groups, fly larvae, midge larvae, and white worms are listed, with potential for fairy shrimp and soldier fly larvae. Among small animals, the examples listed are nematodes, brine shrimp, and water fleas; mosquito larvae and ostracods have potential.

We provide details on some species used to raise larval fish and shellfish: brine shrimp, copepods, and daphnia. In culturing marine species, sometimes both the larval and adult stages of these organisms are used as food, as is the case with brine shrimp. A variety of marine larvae—including barnacles, oysters, and clams—have been used with varying degrees of success in rearing trials of early life stages of marine animals.

Plants

Algae

Unicellular Microalgae

Several genera of microalgae are good food sources for such marine and freshwater animals as clams, rotifers, oysters, and larval fish and shrimp. Hatchery personnel prefer different genera, species, and even strains, and usually have their favorite techniques for rearing them. Six species are used as feed for raising early stages of aquacultural species. Some of these are motile, possessing *flagella,* whiplike structures that help them move about. The number of flagella and their point attachment are useful in species identification. Unicellular microalgae do not have individualized common names, so only their scientific names are used.

One genus of microalgae that is an excellent food source and is popular with aquaculturists is *Isochrysis,* whose Tahitian strain is easy to culture in laboratories and hatcheries. This small, naked flagellate does not have a heavy cell wall, making it desirable for larval stages of herbivores. In high density, *Isochrysis* appear to suppress the growth of bacteria that contaminate the medium.

Tetraselmis spp. are another popular kind of unicellular microalga. They are comparatively large (10–15 μm), green in color (contain chlorophyll), and have flagella located at one end of the oblong cell. They prefer strong light and temperatures as high as 86 °F, and are capable of rapid movement. Aeration may be required to prevent *Tetraselmis* from settling to the bottom and becoming unavailable to the culture species being fed. (Bacteria and protozoans also feed on *Tetraselmis.*) Nutrients, trace metals, and vitamins must be formulated in approved proportions and added to salt water to ensure rapid growth and high survival. The density of individuals is very important: When this reaches high levels, die-off of the population occurs. To keep the population in the *exponential* (rapid-cell-division) *phase,* it is necessary to dilute the culture at intervals or transfer the culture into a larger container with fresh medium. Under good growth conditions, an entire culture can be harvested as early as five days after inoculation of the batch cultures.

One of the most popular kinds of planktonic microalgae used in rearing early stages for many animals is the diatoms. They do not possess obvious flagella

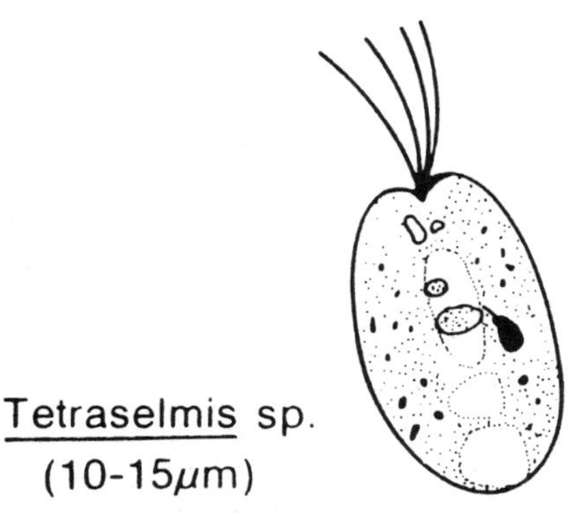

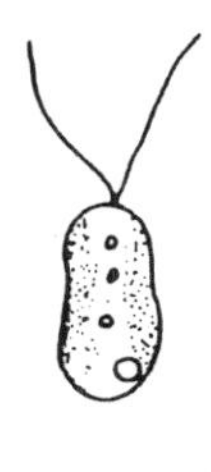

Unicellular algae.

and—despite their glasslike (but perforated), silica cell walls—are of considerable food value. Diatoms are called *primary producers* because they extract inorganic nutrients from the water and convert them to food via photosynthesis. They can be cultivated with relative ease and are of a suitable size to be eaten by small fish and shellfish. They are also consumed by zooplankton, which are then eaten by fish larvae. Procedures for raising diatoms and other microalgae have been established and are available to aquaculturists. Care must be taken not only in rearing them but also in feeding them to animals: It is especially important to determine a suitable ratio of diatoms to number of larvae to be fed. An additional value of diatoms in hatcheries is that they help maintain desirable water quality, since they can use metabolic waste products from larval animals.

A number of useful chemicals can be produced from microalgae by a variety of processes, such as extraction, fermentation (with or without oxygen), and catalytic processing. Since fish and crustaceans obtain Omega-3 (an oil with health benefits to humans) from algae eaten by aquatic species, General Mills, Inc. is using some of the algal lipids experimentally to increase the Omega-3 levels in feeds of farmed fish. *Spirulina* has demonstrated larger weight gains in farmed fish than have several other algal species.

REFERENCES

Fulks, W., and K. L. Main, eds. 1991. *Rotifer and microalgae culture systems; proceedings of a U.S.–Asia workshop*. Waimanalo, HI: Oceanic Institute.

Guillard, R. R. L. 1975. Culture of phytoplankton for feeding marine invertebrates. In *Culture of marine invertebrate animals*, ed. W. L Smith and M. H. Chanley, pp. 29–60. New York: Plenum Press.

Hoff, F. H., and T. W. Snell. 1989. *Plankton culture manual*, 2nd ed. Dade City, FL: Florida Aqua Farms.

Rosen, B. H. 1990. *Microalgae identification for aquaculture*. Dade City, FL: Florida Aqua Farms.

Gracilaria spp.

Macroalgae

Macroalgae, which may be used as human food, can also be raised in tanks and ponds for feeding aquacultural market species: *Gracilaria* spp., for example, are used to feed sea hare. Certain macroalgae provide useful chemicals (discussed below under "Production of Chemicals").

Aquatic Plants

A number of kinds of aquatic plants (duckweeds, azolla, and water hyacinth) are suitable for fish fodder. Difficulties of raising aquatic plants as food for herbivorous fishes include overgrazing on the plants by the fish, clogging of aquaculture systems by rapid plant growth, and harvesting of the fish in ponds with plants.

Terrestrial Plants

Terrestrial plants can also be used to provide certain requirements for fishes. Certain parts of such common vegetables as alfalfa, carrots (leaves), marigolds, and squash may be a good source of protein or vitamins needed by some freshwater fish species.

Invertebrates

Arthropods/Crustaceans

Brine Shrimp

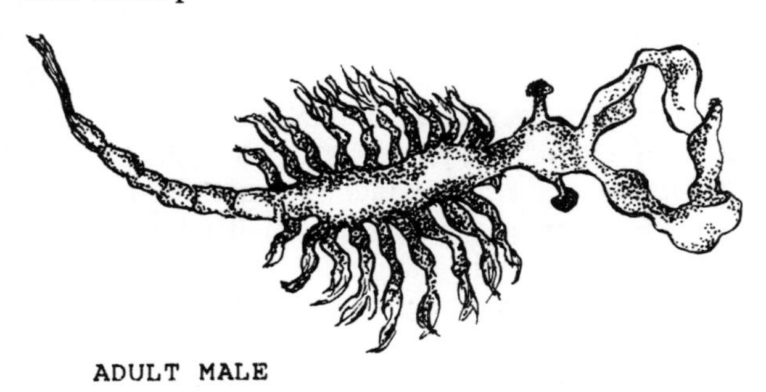

DRIED EGG (CYST)

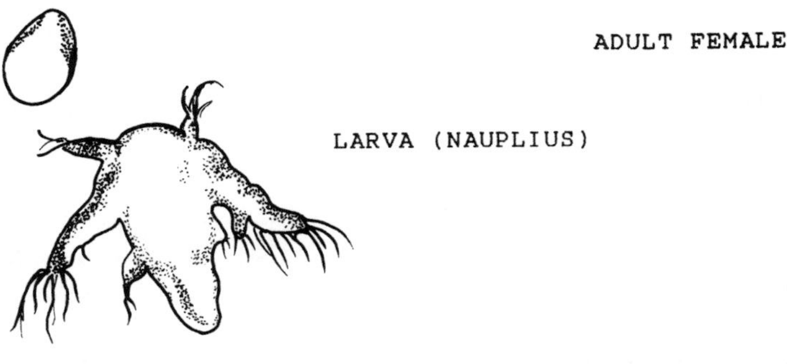

SCIENTIFIC AND COMMON NAMES: ***Artemia salina*, Brine shrimp**

DESCRIPTION AND DISTINCTIVE CHARACTERISTICS: Adult brine shrimp are about one-third of an inch long and have 11 pairs of swimming legs and two stalked, compound eyes. Their variation in color from pale yellow-green to blood red is caused by the concentration and the chemical nature of the salts in solution in ponds and lakes where they live.

HABITAT RANGE: Hypersaline lakes and ponds.

REPRODUCTION AND DEVELOPMENT: Brine shrimp reach sexual maturity within 18–21 days after hatching and reproduce in two ways: If no males are present, the females produce embryos that develop from an unfertilized egg. When both mature females and males are present, the male fertilizes the eggs within the reproductive tract of the female. Eggs remain within the body of the female until a tough shell has formed around each one; then the eggs, which resemble fine brown sand, are deposited in the water, where they will hatch if conditions are favorable. If the female deposits her eggs in sea water with a specific gravity of 1.09 or higher, they will float on the surface and will not hatch; instead, they may be blown by the wind, or carried by water currents to the beaches, where they remain until the rains wash them back into the water. Rain tends to lower the salinity of the surface water, so the eggs usually hatch when they come in contact with it (provided its specific gravity is less than 1.09 and its temperature over 48 ° F). In many instances the eggs hatch within the ovisac of the female, and the embryos develop into *nauplii* (larvae) before birth. The natural environment determines whether a female will bear live young or lay eggs.

AGE AND GROWTH: Only three weeks are required from hatching to maturity. Eggs in nature hatch in great numbers in the spring. Newly hatched brine shrimp have a single eye and one main set of legs, developing the others as they mature into adults. During their growth they acquire a shell, which is shed periodically to allow for further growth.

FOOD AND FEEDING: Brine shrimp eat the free-floating algae and other microscopic organisms that occur in brine ponds and salt lakes. Many diets for culturing *Artemia* have been devised with varying degrees of success. Some experimental diets include different species of unicellular marine algae, yeast, and combinations of the above with various vitamin and chemical additives. Dead algae and bacteria have also been used as food sources for *Artemia* in culture.

PARASITES AND DISEASE: Few reports are available, but viruses, bacteria, fungi, and cestodes have been noted in culture.

PREDATORS AND COMPETITORS: In nature brine shrimp are eaten by saltwater fly larvae, fish, lobsters, some prawns, and birds. Their hypersaline environment excludes some possible predators, such as some of the finfish.

AQUACULTURAL POTENTIAL: Both larval and adult forms of the brine shrimp *A. salina* and related species are widely used as food for many popular marine and freshwater aquarium fish and invertebrates. This tiny shrimp is very much in demand today by fish fanciers and aquaculturists. Because the supply is limited, the few dealers that harvest brine shrimp have greatly increased their price. An unfortunate aspect of brine shrimp is that they may concentrate toxic substances from their environment, including chlorinated hydrocarbons such as DDT; those raised in pollution-free environments would have a distinct market advantage. Facilities for raising brine shrimp can be relatively small.

REGIONS WHERE FARMED AND/OR RESEARCHED: United States (California, Utah), Belgium, Ecuador.

REFERENCES

Browne, R. A., P. Sorgeloos, and C. N. A. Trotman, eds. 1991. Artemia *biology*. Boca Raton, FL: CRC Press.

Helfrich, P., J. Ball, A. Berger, P. Bienfang, S. A. Cattell, M. Foster, G. Fredholm, B. Gallagher, E. Guinther, G. Krasnick, M. Rakowicz, and M. Valencia. 1973. *The feasibility of brine shrimp production on Christmas Island.* Sea Grant Technical Report UNIHI-SEAGRANT-TR-73–02. Honolulu: University of Hawaii.

Lai, L. L. 1991. Production methods and the role of the brine shrimp *Artemia.* In *Production of aquatic animals; crustaceans, molluscs, amphibians and reptiles,* ed. C. E. Nash, pp. 67–77. Amsterdam: Elsevier.

Persoone, G., P. Sorgeloos, O. Roels, and E. Jaspers, eds. 1980. *The brine shrimp* Artemia. Wettern (Belgium): Universal Press.

Sorgeloos, P., P. Lavens, P. Leger, W. Tackaert, and D. Versichele. 1986. *Manual for the culture and use of brine shrimp* Artemia *in aquaculture.* Ghent (Belgium): State University of Ghent, Faculty of Agriculture.

Copepods

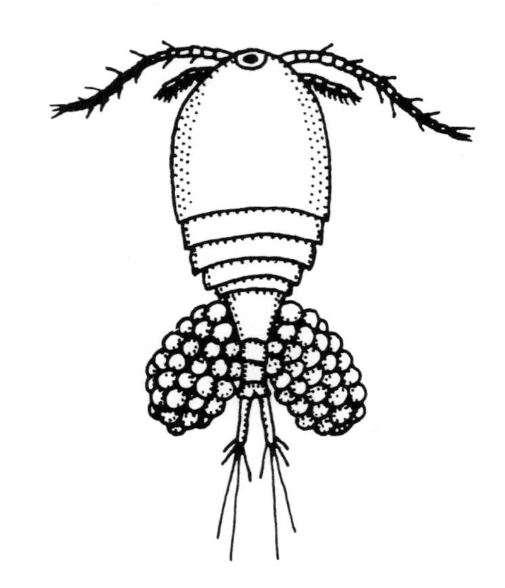

SCIENTIFIC AND COMMON NAMES: ***Acartia tonsa*** **[no common name]** The class Copepoda contains over 4,500 fresh- and saltwater species.

DESCRIPTION AND DISTINCTIVE CHARACTERISTICS: The head of this small crustacean has a single median nauplial eye. The first pair of antennae, large and held at approximately right angles to the bullet-shaped body, are believed to enhance flotation; the second pair are used for swimming. Length is usually less than 3 mm.

HABITAT RANGE: Usually planktonic in the water column in estuaries of the Atlantic Coast of North America. In some areas they become the most abundant zooplankton.

REPRODUCTION AND DEVELOPMENT: Eggs carried in ovisacs attached to the female protect the offspring until they become nauplius larvae, at which time they hatch.

AGE AND GROWTH: *A. tonsa* and other copepods can grow from eggs to adults in about 20 days when held at a temperature of 68 °F.

FOOD AND FEEDING: Filter feeders, eating mostly diatoms. Other copepods are raptorial predators, and still others parasitic.

PARASITES AND DISEASE: They serve as first and second intermediate hosts of fishes, and may be infected with trematodes, cestodes, and nematodes.

PREDATORS AND COMPETITORS: Copepods are very important in the lower levels of the food web, serving as food for a wide variety of larval fish and shellfish; they are also cannibalistic.

AQUACULTURAL POTENTIAL: This and other species of copepod are suitable as food for early stages of marine species. Copepods are also used in pollution bioassays, ecological experiments, and for feeding hatchery fish. They have many desirable characteristics, including relatively short life cycle, high reproductive capacity, the ability to survive in the high densities usually found in culture systems, and good growth under a wide variety of environmental parameters. Various species are widely distributed and occur in high abundance.

REGIONS WHERE FARMED AND/OR RESEARCHED: National Marine Water Quality Laboratory, West Kingston, Rhode Island; Gulf Coast Research Laboratory, Ocean Springs, Mississippi; Marine Biological Institute, University of Texas Medical Branch, Galveston; Danish Institute for Fisheries and Marine Research, Hirtshals, Denmark.

REFERENCES

Kinne, O., ed. 1977. Copepoda. In *Marine ecology*, vol. 3: *Cultivation*, part 2, pp. 761–98. Chichester (U.K.): John Wiley.

Stottrup, J. G., K. Richardson, E. Kirkegaard, and N. J. Pihl. 1986. The cultivation of *Acartia tonsa* Dana for use as a live food source for marine fish larvae. *Aquaculture* 52:87–96.

Uhlig, G. 1984. Progress in mass cultivation of harpacticoid copepods for mariculture purposes. In *Research on aquaculture, Proceedings of the 2nd Seminar of the German–Israeli Cooperation in Aquaculture Research*, ed. H. Rosenthal and S. Sarig, pp. 261–73. European Mariculture Society Special Publication no. 8. Bredene (Belgium): European Mariculture Society.

Water Fleas

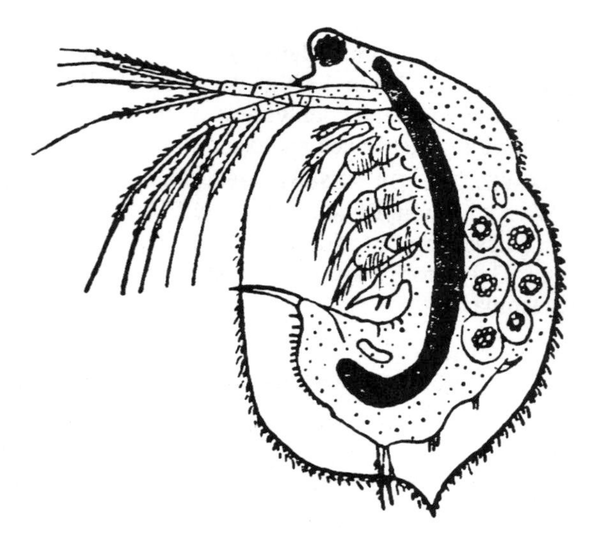

SCIENTIFIC AND COMMON NAMES: ***Daphnia* spp., Water fleas**
Also called daphnia in common parlance. *Moina* spp. are another water flea.

DESCRIPTION AND DISTINCTIVE CHARACTERISTICS: Daphnia are small crustaceans, about 0.75 in. when adults. The *carapace* (external "shell") encloses five pairs of legs. *Biramous* (two-branched) second antennae extend from the animal's "head," which also bears a pair of compound eyes.

HABITAT RANGE: Found in stagnant waters of lakes and ephemeral ponds, ditches, swamps, and so on, where decomposing organic material is present. Different genera and species are found worldwide.

REPRODUCTION AND DEVELOPMENT: A female may produce over a hundred eggs every two or three days. When environmental conditions are adverse (low oxygen, lack of food, high density), resting eggs are released to survive until conditions improve; this is a sexual stage. Under suitable conditions, reproduction is asexual, with females producing more females, and no males required. Some species can reproduce when only four to seven days old.

AGE AND GROWTH: The carapace must be shed to allow for growth, which is rapid.

FOOD AND FEEDING: In the wild, daphnia eat bacteria, yeast, and microalgae. Nonliving materials, such as detritus and dissolved organic matter, also serve as food.

PARASITES AND DISEASE: There is no record of parasites or disease in wild specimens. In culture, death may result when the medium holding daphnia is changed rapidly, since the new medium may not be "conditioned" and may contain "toxic substances." Failure of daphnia in cultures to molt completely (with the old exoskeleton clinging to the animal) causes death. A yeast causes hemocoelic infection.

PREDATORS AND COMPETITORS: Water fleas are an important source of food for larval fish. Rotifers, ciliates, and copepods compete with daphnia for food.

AQUACULTURAL POTENTIAL: For over a century water fleas have been used as live food for fish in hatcheries. They are desirable food for newly hatched fry, such as goldfish and freshwater tropical fish. Their protein content based on dry weight is roughly 50 percent.

REGIONS WHERE FARMED AND/OR RESEARCHED: Florida, California.

REFERENCES

Hoff, F. H., and T. W. Snell. 1989. *Plankton culture manual,* 2nd ed. Dade City, FL: Florida Aqua Farms.

Rees, J. T., and J. M. Oldfather. 1980. Small-scale mass culture of *Daphnia magna* Straus. *Proceedings of the World Mariculture Society* 11:202–10.

Rotifers

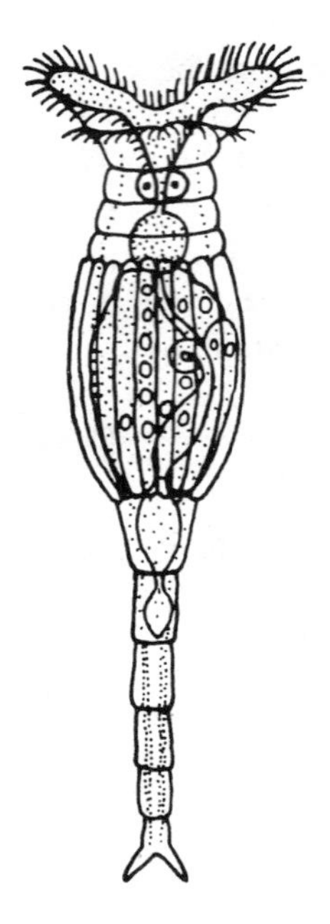

SCIENTIFIC AND COMMON NAMES: ***Brachionus plicatilis*, Rotifer**
Also called wheel animalcules. Some zoologists believe rotifers belong in a phylum by themselves (Rotifera); others that they should be placed in the phylum Aschelminthes.

DESCRIPTION AND DISTINCTIVE CHARACTERISTICS: Body consists of a head, trunk, and an elongated body. A ciliated band surrounds the ciliated *corona* (crown). Total length is usually less than 400 μm.

HABITAT RANGE: Cosmopolitan, euryhaline (1–60 ppt salinity) species, living in the water column in water temperatures of 68–86 °F; apparently survives better in brackish water.

REPRODUCTION AND DEVELOPMENT: Very high reproductive rate: Females mature in a few days and males impregnate them soon thereafter. Eggs are released every 4–6 hr, hatching in about 12 days. A female may produce about 25 young during her short life.

AGE AND GROWTH: The life span of a rotifer is very short: Males may only live two days, but females as long as eight.

FOOD AND FEEDING: Herbivores, eating the unicellular green algae *Chlorella,* which works well in culture; yeast is also used to culture rotifers.

PARASITES AND DISEASE: Bacteria; cysts can be treated with antibiotics to reduce bacterial diseases in rearing facilities.

PREDATORS AND COMPETITORS: Eaten by a wide variety of larval fish.

AQUACULTURAL POTENTIAL: The role of rotifers as food for larval fish was developed in Japan in the early 1960s; they are now important as first food for larval fish in hatcheries worldwide. Rotifers have been likened to a "brown lunch bag," wherein highly unsaturated fatty acids, when fed to rotifers just before feeding them to larval fish, produce greater larval fish survival. Though not itself highly nutritious, with added fatty acids the rotifer becomes a live food item that offers a good diet. A problem of mass cultures of rotifers is that they are unstable. Depression of rotifer swimming activity suggests that there is a need for improved water quality or a change in feeding procedures.

REGIONS WHERE FARMED AND/OR RESEARCHED: United States, Japan, Britain.

REFERENCES

Fontaine, C. T., and D. B. Revera. 1980. The mass culture of the rotifer, *Brachionus plicatilis,* for use as foodstuff in aquaculture. *Proceedings of the World Mariculture Society* 11: 211–18.

Hoff, F. H., and T. W. Snell. 1989. *Plankton culture manual,* 2nd ed. Dade City, FL: Florida Aqua Farms.

Lubzens, E. 1987. Raising rotifers for use in aquaculture. *Hydrobiologia* 147:245–55.

Annelids/Aquatic Oligochaetes

Tubifex Worms

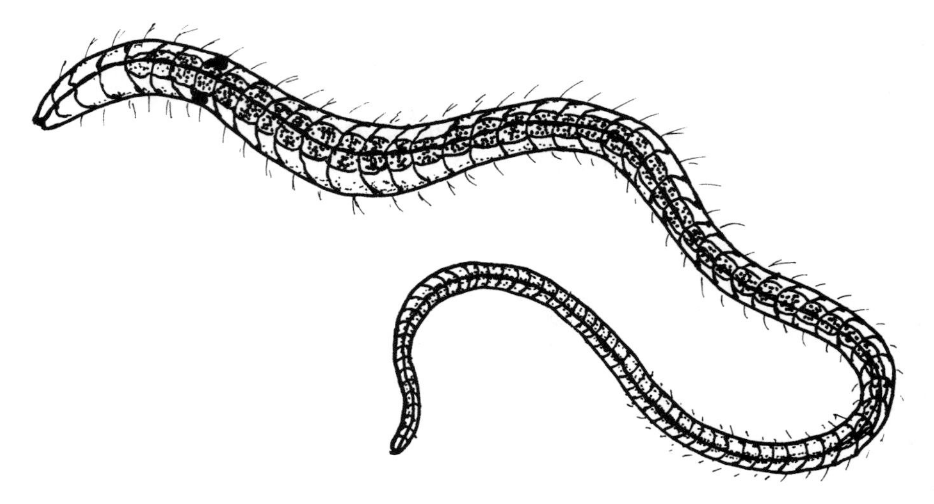

SCIENTIFIC AND COMMON NAMES: ***Tubifex tubifex*, Tubifex worms**
Related: *Limnodrilus* spp., *Branchiura sowerbyi*, *Lumbriculus variegatus*

DESCRIPTION AND DISTINCTIVE CHARACTERISTICS: These worms are about 2 in. in length, thin, and red in color.

HABITAT RANGE: Live in muddy shores with low oxygen levels.

REPRODUCTION AND DEVELOPMENT: Females become mature at about age 3 months, and lay an egg capsule containing as many as 200 larvae every 7–10 days.

AGE AND GROWTH: Populations double every 11–42 days.

FOOD AND FEEDING: Tubifex worms consume decaying organic matter.

PARASITES AND DISEASE: Aquatic oligochaetes, such as tubifex worms, may carry certain bacterial diseases harmful to fish. A protozoan flagellate infects these worms causing pigmentation loss, reduced activity, and death. Tubifex worms serve as intermediate hosts for a myxosporean that causes "whirling disease" in trout that eat the worms. Heavy infections of cestodes can cause high mortality in tubifex worms.

PREDATORS AND COMPETITORS: Tubifex worms serve as food for a variety of invertebrates used in research (e.g., planarians, leeches, dragonfly and damselfly nymphs), as well as for finfish.

AQUACULTURAL POTENTIAL: Species of salmonids and sturgeons have shown increased appetites and palatability, as well as relatively high growth and survival rates, when fed this live food source. Tubifex worms have also been used to feed frogs and prawns. The cost of producing these live feeds is comparable to costs of manufacturing inanimate (pelleted) feeds, and their use can reduce pollution in aquaculture facilities. There is a small demand for live worms at biological supply houses.

REGIONS WHERE FARMED AND/OR RESEARCHED: Bogotá, Cali, and Medellín, Colombia; Cuba.

REFERENCES

Brinkhurst, R. O., and D. G. Cook, eds. 1980. *Aquatic oligochaete biology*. New York: Plenum Publishing Corp.

Lietz, D. M. 1987. Potential for aquatic oligochaetes as live food in commercial aquaculture. *Hydrobiologia* 155:309–10.

Marian, M. P., and T. J. Pandian. 1984. Culture and harvesting techniques for *Tubifex tubifex*. *Aquaculture* 42:303–15.

SPORTFISH BAIT

The worldwide popularity of recreational fishing has grown immensely in recent years. In 1985, expenditures associated with angling in the United States averaged $604 per adult sport angler, for a U.S. total of over $28 billion. This increase in sportfishing produced a strong demand for bait. A wide variety is sold, including both artificial bait and natural, terrestrial animals (e.g., insects, worms); however, demand has stimulated the aquaculture of marine bait worms (bloodworms and lugworms), as well as that of bait fish and shellfish (shrimp) from both salt and fresh water.

The kinds of animals raised for this market of necessity must be "desired or preferred" food of the fish sought by the anglers. They also must be "angler friendly," withstand sorting, and survive transport in high-density containers. (Bait farming operations require considerable handling and grading of fish and the shipping of live animals.) In general bait animals are small, numerous, and frequently heavily parasitized: Characteristically they are prey species serving as intermediate hosts to a variety of parasites.

Invertebrates

Annelids/Marine Bait Worms

Bloodworms

SCIENTIFIC AND COMMON NAMES: ***Glycera dibranchiata*, Bloodworm,** Related: sandworm, *Nereis* spp.

DESCRIPTION AND DISTINCTIVE CHARACTERISTICS: Similar in appearance to the common earthworm. Bloodworms have a red-colored body fluid that contains a respiratory pigment similar to hemoglobin.

HABITAT RANGE: Intertidal zone from the Canadian Maritimes to New England, in areas where bottom sediments are stable enough to support their burrows.

REPRODUCTION AND DEVELOPMENT: Females release 1.2–3 million eggs during the summer; these hatch into planktonic larvae.

AGE AND GROWTH: Bloodworms are 8–14 in. long by their third year of life.

FOOD AND FEEDING: Feed at night upon dead and decaying plant materials.

PARASITES AND DISEASE: Some species are parasitized by gregarines (protozoans).

PREDATORS AND COMPETITORS: Used as bait for blackfish, bluefish, flounder, fluke, kingfish, porgy, sea bass, striped bass, and weakfish, chiefly on the U.S. East Coast between New York and North Carolina. The suitability of these worms as bait suggests that these fish prey on them in nature. They are also eaten by herring gulls.

AQUACULTURAL POTENTIAL: The high demand for bait species and the rapid growth and feeding habits of this worm suggest rearing can be profitable, although demand is seasonal and may be somewhat limited geographically. Worms can be shipped long distances inexpensively.

REGIONS WHERE FARMED AND/OR RESEARCHED: Canada.

REFERENCES

Klawe, W. L., and L. M. Dickie. 1957. *Biology of the bloodworm,* Glycera dibranchiata *Ehlers, and its relation to the bloodworm fishery of the Maritime Provinces.* Bulletin of the Fisheries Research Board of Canada no. 115. Ottawa: Fisheries Research Board of Canada.

Pope, E. C. 1965. Can marine worms be farmed? *Australian Fisheries Newsletter* 24(2):13–15.

Lugworms

SCIENTIFIC AND COMMON NAMES: ***Arenicola cristata,* Lugworm**

DESCRIPTION AND DISTINCTIVE CHARACTERISTICS: Earthwormlike animals.

HABITAT RANGE: Dwell in tubes in the intertidal zone of both the Atlantic and Pacific oceans, in waters with a summer temperature of about 68 °F.

REPRODUCTION AND DEVELOPMENT: Females spawn eggs in a mucus mass year-round; larvae settle close to where eggs are laid. Lugworms mature within two or three months.

AGE AND GROWTH: Market size reached in about three months.

FOOD AND FEEDING: Eat decaying marine plants and microorganisms, such as fungi, bacteria, and protozoans.

PARASITES AND DISEASE: Some protozoans are probably parasitic.

PREDATORS AND COMPETITORS: In nature, lugworms are consumed by many finfish species.

AQUACULTURAL POTENTIAL: The U.S. bait market is strong and getting stronger, but demand is seasonal. Generally, lugworm aquaculture would be a small operation requiring large volume and netting a small profit margin. If collecting lugworms from state waters is prohibited, the chances of private culturing becoming profitable are improved.

REGIONS WHERE FARMED AND/OR RESEARCHED: Florida.

REFERENCE

D'Asaro, C. N., and H. C. K. Chen. 1976. *Lugworm aquaculture.* Florida Sea Grant Program Report no. 16. Gainesville: Florida Sea Grant College Program.

Arthropods/Crustaceans

Grass Shrimp

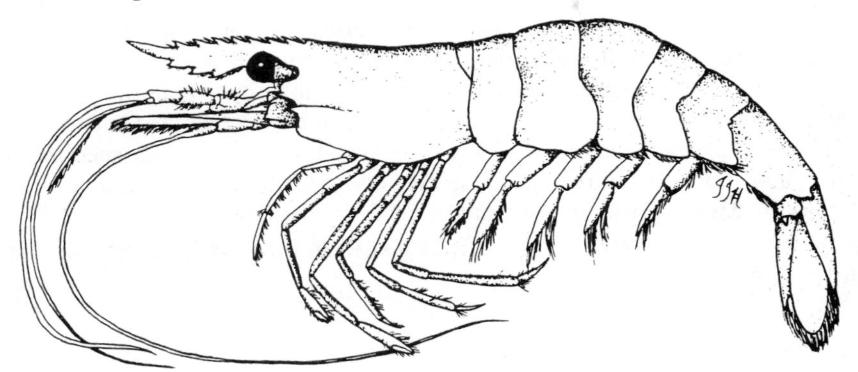

SCIENTIFIC AND COMMON NAMES: ***Palaemonetes pugio*, Grass shrimp** Related: *P. vulgaris, P. intermedius.* Also called glass or ghost shrimp.

DESCRIPTION AND DISTINCTIVE CHARACTERISTICS: Small shrimp, about 2 in. long when mature. They are almost transparent (hence the variant name "glass shrimp"), and resemble small *Penaeus* shrimp, but are a bit more robust.

HABITAT RANGE: There are many species of small *Palaemonetes* shrimp that can reach extreme abundances in shallow fresh-, brackish-, and salt-water habitats with grassy bottoms (hence the name "grass shrimp"). Various species (there are at least five in Louisiana alone) tolerate various salinities. *P. pugio* is a common estuarine species.

REPRODUCTION AND DEVELOPMENT: In contrast to penaeid shrimps, which release their eggs free in the water, mature grass shrimps carry their eggs on their abdomens. Salt-water species release 100–400 small eggs per spawning, and a single female probably spawns more than once each year.

AGE AND GROWTH: Grass shrimp may reach adult size in two to four months even in high-density situations.

FOOD AND FEEDING: Omnivorous, feeding on algae, zooplankton, meiofauna, and detritus.

PARASITES AND DISEASE: The health of laboratory-reared grass shrimp declines the longer they are maintained: Two weeks or so seems to be the limit for healthy animals. Protozoans (ciliates and microsporeans) and isopods parasitize grass shrimp, and chinoclastic bacteria are thought to be possible pathogens.

PREDATORS AND COMPETITORS: Many species of fish and birds.

AQUACULTURAL POTENTIAL: Grass shrimp make ideal bait for a variety of fish species, including bream, crappie, catfish, and striped and largemouth bass. They are a fine natural food for rearing game fish and can tolerate densities as high as 500,000–2,500,000 per acre of water. Because they are rather easily obtained in nature, it is probable that no one farms them on a moderate or large scale, although a small-scale operation can be managed in conjunction with some other type of aquaculture. Their wide use in toxicity testing (metallic, organic, and petroleum contaminants) has resulted in raising them for sale to commercial suppliers to this market; they are also sold as aquarium specimens. Their aquaculture requires only a small pond with bottom vegetation and a means of keeping predators (fish and birds) from the feast.

REGIONS WHERE FARMED AND/OR RESEARCHED: Maryland, Louisiana, Mississippi, Alabama, and Georgia.

REFERENCES

Anderson, G. 1985. *Species profiles: Life histories and environmental requirements of coastal fishes and invertebrates (Gulf of Mexico)—Grass shrimp.* Biological Report 82(11.35). Slidell, LA: U.S. Fish and Wildlife Service, National Coastal Ecosystems Team.

Buikema, A. L., Jr., B. R. Niederlehner, and J. Cairns, Jr. 1980. Use of grass shrimp in toxicity tests. In *Aquatic invertebrate bioassays,* ed. A. L. Buikema, Jr., and J. Cairns, Jr., pp. 155–73. ASTM Special Technical Publication no. 715. Philadelphia: American Society for Testing and Materials.

Hoff, F. H., and T. W. Snell. 1989. *Plankton culture manual,* 2nd ed. Dade City, FL: Florida Aqua Farms.

Huner, J. V., and E. E. Brown, eds. 1985. *Crustacean and mollusk aquaculture in the United States.* Westport, CT: AVI Publishing Co.

McLarney, W. 1984. *The freshwater aquaculture book; a handbook for small scale fish culture in North America.* Point Roberts, WA: Hartley & Marks.

Vertebrates/*Bait Fish*

There is considerable biological information available on bait fish. Some states encourage their being farmed because fishing on natural stocks of small fish for bait may include the young of commercial and recreational fish. Heavy fishing by anglers on small fish for bait may also reduce the supply of forage fish available to recreational fish species. Most farmed bait fish are raised in fresh water.

Some states prohibit the sale of certain bait fish (e.g., goldfish and tilapia). A partial list of species sold in the United States follows:

Chub suckers, *Erimyzon* spp.
Fathead minnow, *Pimephales promelas*
Goldfish, *Carassius auratus*
Gulf killifish, *Fundulus grandis*
Killifish, *Fundulus* spp.
Shiners, *Notropis* spp.
Stone rollers, *Campostoma* spp.
Tilapia, *Tilapia* spp.
Top minnows, *Poecilia* spp.

Most of these species are from fresh water; the exception is the Gulf killifish, which is commonly found in brackish waters and is discussed below. We also describe the golden shiner, one of the most important bait fishes and one that is similar to many of the other kinds that are sold. (See below under "Ornamental Species" for goldfish, and Chapter 3 under "Finfishes" regarding tilapia.)

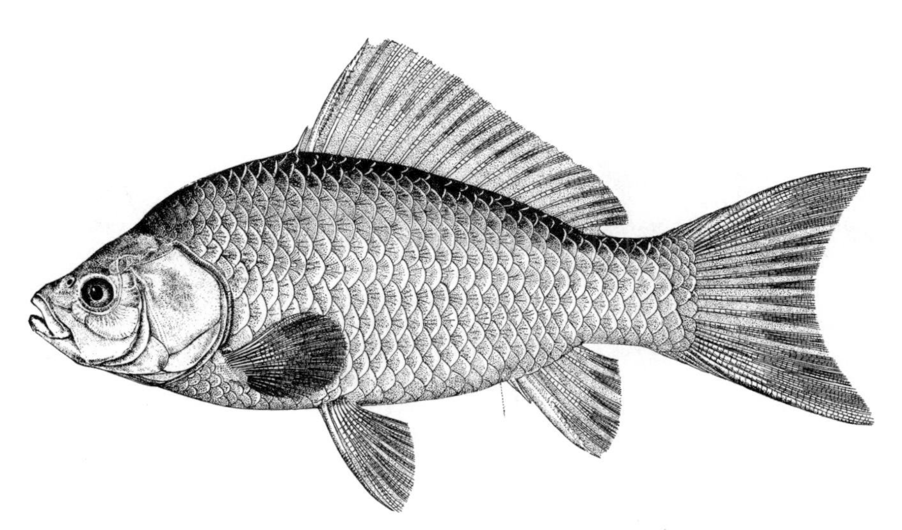

Goldfish (*Carassius auratus*), nonornamental variety.

REFERENCES

Giudice, J. J., D. L. Gray, and J. M. Martin. 1982. *Manual for bait fish culture in the South.* Little Rock: University of Arkansas, Cooperative Extension Service.

Stickney, R. R., ed. 1986. *Culture of nonsalmonid freshwater fishes.* Boca Raton, FL: CRC Press.

Strawn, K., P. W. Perschbacher, R. Nailon, and G. Chamberlain. 1986. *Raising mudminnows.* Report of the Texas A&M University Sea Grant Program TAMU-SG-86–506R. Galveston: Texas A&M Sea Grant College Program.

Tatum, W. M., J. P. Hawke, R. V. Minton, and W. C. Trimble. 1982. *Production of bull minnows* (Fundulus grandis) *for the live bait market in coastal Alabama.* Alabama Marine Resources Bulletin no. 13. Dauphin Island: Alabama Marine Resources Laboratory.

Golden Shiner

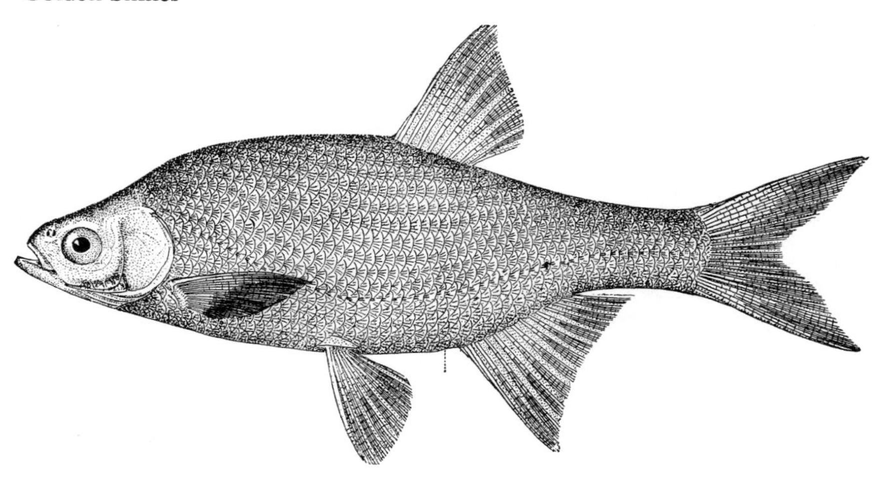

SCIENTIFIC AND COMMON NAMES: ***Notemigonus crysoleucas,*** **Golden shiner**

DESCRIPTION AND DISTINCTIVE CHARACTERISTICS: A bright, colorful, deep-bodied fish with laterally compressed body. Some strains have orange or red fins.

HABITAT RANGE: Large streams and lakes throughout the central and eastern United States.

REPRODUCTION AND DEVELOPMENT: Spawning takes place at temperatures of 70 °F or above, usually during May–August in some parts of the golden shiner's range, with females depositing up to 10,000 eggs on vegetation. They become sexually mature at about age 1 year.

AGE AND GROWTH: Grows from egg to about 4 in. in a single summer.

FOOD AND FEEDING: Feeds on aquatic vegetation, especially filamentous algae.

PARASITES AND DISEASE: This species has numerous parasites and is usually heavily parasitized. One protozoan parasite of ovaries causes mortality; tapeworms, leeches, and copepods are also present.

PREDATORS AND COMPETITORS: Predators include turtles, birds, and larger fish.

AQUACULTURAL POTENTIAL: Production of this bait fish is greatly affected by water chemistry. Profitable management requires that ponds be productive at all times since artificial feeding is not practiced. Production of about 600–800 lb per acre is not uncommon: One large farm in Arkansas with over 7,000 acres in golden shiners sells to three-fourths of the United States (Arizona to New York). Since they are shipped alive, care must be taken that they are in good condition, and the considerable hand labor required in counting and sorting by size and quality makes the product more expensive. Demand by recreational fishermen is seasonal. Parasites, diseases, predation, and pond flooding after heavy rains can severely reduce production.

REGIONS WHERE FARMED AND/OR RESEARCHED: Arkansas, Mississippi, Missouri, Louisiana, Texas, and Alabama.

REFERENCES

Dobie, J. R., O. L. Meehean, and G. N. Washburn. 1948. *Propagation of minnows and other bait species.* U.S. Fish and Wildlife Service Circular no. 12. Washington, DC: U.S. Government Printing Office.

Giudice, J. J., D. L. Gray, and J. M. Martin. 1982. *Manual for bait fish culture in the South.* Little Rock: University of Arkansas, Cooperative Extension Service. 49 pp.

Minnow farming is big business. 1977. *Commercial Fish Farmer and Aquatic News* 3(4):4–9.

Stickney, R. R., ed. 1986. *Culture of nonsalmonid freshwater fishes.* Boca Raton, FL: CRC Press.

Gulf Killifish

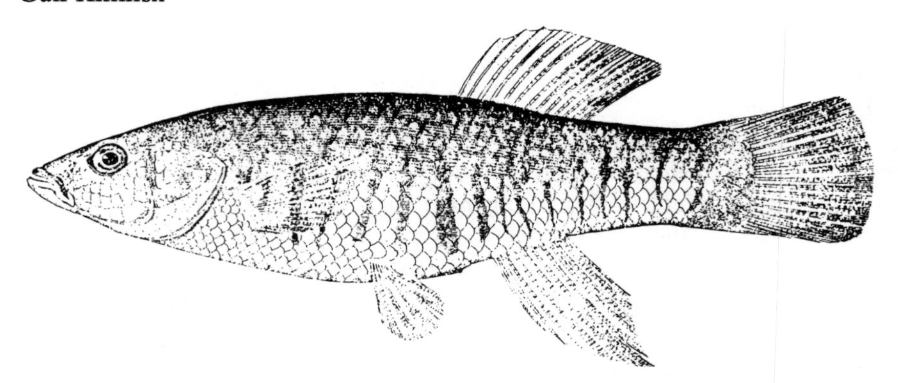

SCIENTIFIC AND COMMON NAMES: ***Fundulus grandis,* Gulf killifish** Also called the mud fish, mud minnow, and bull minnow.

DESCRIPTION AND DISTINCTIVE CHARACTERISTICS: A small fish (<6 in.) with fins devoid of spiny rays. Females are uniformly greenish-silver; males are darker and have prominent spots.

HABITAT RANGE: Along shores and in estuaries in the St. John's River and down along Florida's Atlantic Coast through the northern Gulf of Mexico to Veracruz, Mexico.

REPRODUCTION AND DEVELOPMENT: Gulf killifish spawn when the water temperature reaches 68 °F; times may differ according to location. Spawning may last all year or, when conditions are favorable, be limited to one or two periods. Spawning mats are made of Spanish moss placed between two sheets of plastic coated wire.

AGE AND GROWTH: Rate of growth is dependent on pond density. Four or five months are required for Gulf killifish to reach a length of 2.5 in.

FOOD AND FEEDING: Gulf killifish adapt well to a variety of commercial fish feeds, such as minnow meal and catfish pellets. Agricultural by-products, such as cottonseed meal and wheat shorts, are inexpensive and suitable for rearing fry. Natural food consists of zooplankton, aquatic worms, and insect larvae.

PARASITES AND DISEASE: These fish can tolerate high levels of infection of both parasites and disease organisms. Protozoan parasites (sporozoans, coccidians) infect bull minnows in the wild. Four different species of trematodes, one species of acanthocephalan crustacean, and bacteria occur in Gulf killifish.

PREDATORS AND COMPETITORS: Eaten by a wide range of large fish and piscivorous birds. At high stocking densities, growth and survival may be reduced.

AQUACULTURAL POTENTIAL: A hardy fish that adapts well to captivity and is popular as bait along much of the northern Gulf Coast from New Orleans to Pensacola (FL). Their tolerance to high salinity makes them among the few fish species that can be cultured for the marine bait market. The marketing aspects of this species and especially the seasonal demand (which can be very low in winter) should be investigated thoroughly before considering investment. Gulf killifish farmers must compete with wild-caught production, as well as with bait shrimp.

REGIONS WHERE FARMED AND/OR RESEARCHED: Texas, Mississippi, and Alabama.

REFERENCES

Giudice, J. J., D. L. Gray, and J. M. Martin. 1982. *Manual for bait fish culture in the South.* Little Rock: University of Arkansas, Cooperative Extension Service.

Stickney, R. R., ed. 1986. *Culture of nonsalmonid freshwater fishes.* Boca Raton, FL: CRC Press.

Strawn, K., P. W. Perschbacher, R. Nailon, and G. Chamberlain. 1986. *Raising mudminnows.* Report of the Texas A&M University Sea Grant Program TAMU-SG-86–506R. Galveston: Texas A&M Sea Grant College Program.

Tatum, W. M., J. P. Hawke, R. V. Minton, and W. C. Trimble. 1982. *Production of bull minnows* (Fundulus grandis) *for the live bait market in coastal Alabama.* Alabama Marine Resources Bulletin no. 13. Dauphin Island: Alabama Marine Resources Laboratory.

Tatum, W. M., J. P. Hawke, R. V. Minton, and W. C. Trimble. 1982. Alabama researchers study Gulf killifish: Prime culture candidate in coastal regions. *Aquaculture Magazine* 8(4): 20–5.

Wass, B. P., K. Strawn, M. Johns, and W. Griffin. 1983. Commercial production of mudminnows (*Fundulus grandis*) for live bait: A preliminary economic analysis. *Texas Journal of Science* 35(1):51–60.

EXPERIMENTAL ANIMALS

BIOASSAY AND AQUATIC TOXICOLOGY SPECIES

Although the two terms are frequently interchanged, there is a difference between a *bioassay* and a *toxicity test,* as pointed out by Rand and Petrocelli (1984). "Aquatic toxicology has been defined as the study of the effects of chemicals and other foreign agents on aquatic organisms with special emphasis on adverse or harmful effects ... Aquatic toxicity tests are used to detect and evaluate the potential toxicological effects of chemicals on aquatic organisms."

A bioassay, on the other hand, is defined by these authors as "a test to evaluate the relative potency of a chemical by comparing its effect on a living organism with that of a standard preparation ... [and] *not* to estimate the concentration of the chemical that is *toxic* to those organisms" (emphasis added). Hence, bioassays simply determine a chemical's strength, based on the degree of response elicited in the test organisms. The testing of pharmacologically active compounds, such as vitamins, is done using bioassays. With the great emphasis on environmental pollution, prompted by the proliferation of humankind's waste products, toxic studies have become very important to state and federal environmental protection agencies.

For toxicity studies and bioassays to be useful, it is imperative that the test and control animals are identical in all ways possible, including age, size, and genetic makeup. They must also be abundant enough to provide sufficiently large sample sizes for statistical proof; be healthy (i.e., not infected or infested by parasites); consume food during the tests; and be handled and shipped over rather long distances with minimum mortality. Test animals raised in outdoor ponds or caught in the wild do not meet these strict requirements; it is those raised in aquaria in highly controlled conditions that are best suited for such studies.

A considerable fund of knowledge concerning the rearing and maintaining of species used in bioassays and toxicity tests was a valuable by-product of literally hundreds of years of aquacultural experience. Aquaculture researchers, commercial fish farmers, and fish hobbyists working with invertebrates and vertebrates have greatly expedited bioassay and toxicological testing.

Such a wide variety of species are used in toxicological studies that we cannot list them here; some are covered in the preceding sections (copepods, daphnia, grass shrimp, and bait fish). They include many species of invertebrates (phytoplankton, protozoans, coelenterates, annelids, crustaceans, insects, and mollusks) as well as fishes that live in fresh-, sea-, and brackish-water habitats. The small size and short life cycles of many aquatic invertebrates make them practical for use in bioassay research. Despite the progress made in rearing and maintaining certain species, culture methods have not been determined for many other species, and the variation in toxicant sensitivity of some species over their life cycles is a deterrent to their use. Means of determining cause of death is also a problem in some species.

A word about the market for bioassay and toxicological test animals: While the price paid by laboratories for individual live specimens is usually high, the market is relatively small and may fluctuate considerably with the availability of federal and state funding for environmental projects.

REFERENCES

Buikema, A. L., Jr., and J. Cairnes, Jr. 1980. *Aquatic invertebrate bioassays: A symposium.* ASTM Special Technical Publication no. 715. Philadelphia: American Society for Testing and Materials.

Rand, G. M., and S. R. Petrocelli, eds. 1984. *Fundamentals of aquatic toxicology.* Washington, DC: Hemisphere Publishing Corp.

Franson, M. A. H., ed. 1989. *Standard methods for the examination of water and wastewater,* 17th ed. Washington, DC: American Public Health Association.

BIOMEDICAL ASSAY SPECIES

Several of the species included in this book are useful for biomedical purposes, serving as aquatic "white mice" in experiments involving organ function, disease, and other physiological aspects, eventually yielding a better understanding of human bodily functions and malfunctions. These species have certain biological attributes that are desired by researchers, whether intelligence (as in octopuses) or giant nerve fibers (as in the squid, *Loligo* spp.). The simple nervous system of the sea hare (*Aplysia californica*) is used in many biomedical fields. Some biomedical (and bioassay) species are also useful as teaching aids in school and college biological laboratories.

In studies of disease, such as cancerous tumors, the test animals themselves must be susceptible: Often this is observed in feral fishes collected from polluted environments. Many different chemicals induce tumors in these fish, some as rapidly as a few months after exposure. Study of induced tumors in fish helps medical researchers understand the causes of tumors in humans.

For obvious reasons, small fish are generally desirable for laboratory research, including guppies (*Poecilia reticulata*), zebra fish (*Brachydanio rerio*), and mollies (*Poecilia formosa*). Experiments involving experimental manipulation of body parts or functions are facilitated using larger fish, such as rainbow trout (*Oncorhynchus mykiss*) and carp (*Cyprinus carpio*).

The requirements for species in the biomedical research market are generally the same as those for bioassay/toxicological test species: small, healthy animals with a known history of development, available in large numbers, and easily shipped with high survival. As for bioassay species, the price paid for individual live specimens is high, but the market relatively small and tied in with federal and state research funds.

REFERENCE

U. S. Department of Health and Human Services. 1984. *Use of small fish species in carcinogenicity testing*. National Cancer Institute Monograph no. 65. Bethesda, MD: National Cancer Institute.

Mollusks

Gastropods

Sea Hare

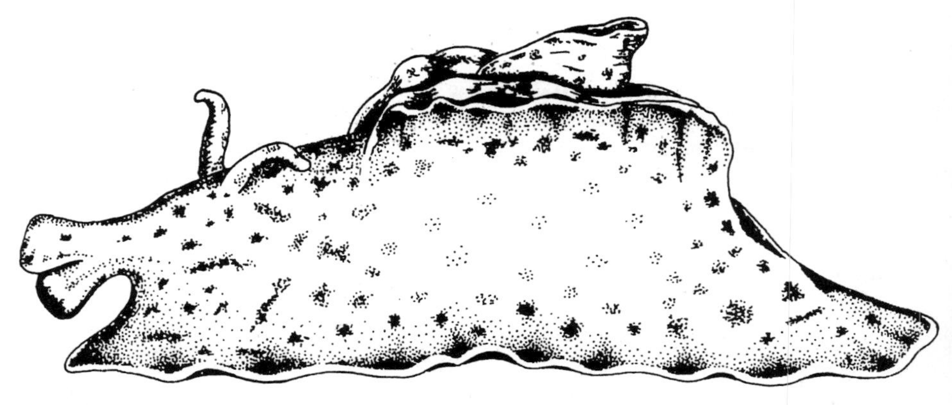

SCIENTIFIC AND COMMON NAMES: ***Aplysia californica*, Sea hare**
Also called sea slug; formerly *Tethlys californica*.

DESCRIPTION AND DISTINCTIVE CHARACTERISTICS: This large mollusk, which resembles a land slug, has an internal shell buried in its mantle. It has four tentacles on its anterior end, with the eyes located posterior of the anterior pair. Large flaps extend from either side of this bulky animal, which moves or crawls about on a slimy foot. Color varies from greenish to reddish-brown to black.

HABITAT RANGE: Abundant in rock tide pools from Monterey to San Diego, California.

REPRODUCTION AND DEVELOPMENT: Eggs are fertilized internally and laid in gelatinous masses; a large female may release seven million. Upon hatching the planktonic veliger larvae swim for approximately one month, after which they settle on red algae and metamorphose into benthic juveniles.

AGE AND GROWTH: Usually reach full size (up to 15 in.) in about a year.

FOOD AND FEEDING: Sea hares are grazing herbivores that may feed on many different varieties of algae.

PARASITES AND DISEASE: Bacteria and protozoans can cause mortalities in cultures.

PREDATORS AND COMPETITORS: Starfish, tulip shell, and jellyfish may be predators in nature, based on laboratory experiments. If disturbed, it releases a heavy, dark purplish fluid that conceals its presence and either frightens a potential predator, or perhaps warns that it is toxic.

AQUACULTURAL POTENTIAL: Sea hares are used rather widely for research in a variety of biomedical fields, such as neurobiology, behavioral science, the hormonal control of reproductive rhythms, and developmental biology. Their entire life cycle can be carried out indoors in a relatively small area, and good quality cultured specimens command a good price. A disadvantage is that sizable quantities of live algae must be grown and maintained for feeding to the juveniles and adults. A major problem that existed for a long time was that cultured veligers tended to become trapped at the air–water interface; however, this has now been solved by decreasing the surface film with cetyl alcohol.

REGIONS WHERE FARMED AND/OR RESEARCHED: University of Miami, RSMAS.

REFERENCES

Carefoot, T. H. 1987. *Aplysia:* Its biology and ecology. *Oceanography and Marine Biology: An Annual Review* 25:167–284.

Switzer-Dunlap, M., and M. G. Hadfield. 1981. Laboratory culture of *Aplysia*. In *Marine invertebrates; laboratory animal management,* pp. 199–216. Washington, DC: National Academy Press.

Cephalopods

Octopus

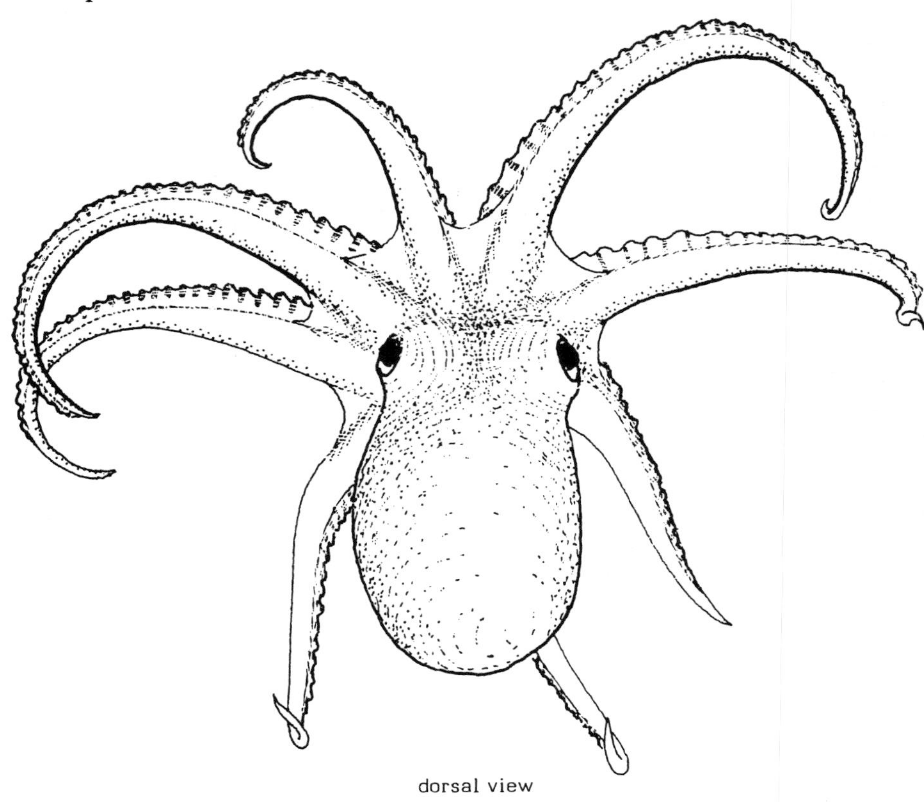

dorsal view

SCIENTIFIC AND COMMON NAMES: ***Octopus joubini,*** **Pygmy octopus**

DESCRIPTION AND DISTINCTIVE CHARACTERISTICS: Short, saclike body with eight arms, one of which (*hectocotylus*) is specialized for use in reproduction. There are no lateral fins. Like all octopuses, *P. joubini* has the ability to change color rapidly over a wide range of colors: red, brown, green, or white-spotted.

HABITAT RANGE: Shallow waters in the tropical western Atlantic: Florida, the Gulf of Mexico, and the Caribbean Sea southward to French Guiana.

REPRODUCTION AND DEVELOPMENT: Females become sexually mature at roughly 20 weeks of age and lay up to about 300 eggs. The pygmy and many other octopus species produce large eggs and large hatchlings that have no larval stage; small-egg species have never been cultured successfully.

AGE AND GROWTH: Under suitable conditions, two crops can be produced annually.

FOOD AND FEEDING: Pygmy octopuses are carnivores and important predators on numerous species of live animals, including clams, snails, crustaceans (shrimps and crabs), and fish. They also eat other octopuses and, occasionally, worms. They may not accept artificial or dead animals as food, though conditioning experiments are considered to overcome this difficulty, which arises in maintaining young octopuses in culture.

PARASITES AND DISEASE: Mesozoans are parasitical; there are unconfirmed reports of cestodes. Bacterial (*Vibrio*) ulcers in high-density cultures have been noted and resulted in high mortalities.

PREDATORS AND COMPETITORS: Octopus are eaten by a variety of crabs, fishes, seals, whales, and birds. In culture they can be cannibalistic.

AQUACULTURAL POTENTIAL: This species has attributes of interest to aquaculturists: It is hardy and can be raised in the laboratory (it has a direct life history). Pygmy octopuses can be maintained if provided with a large volume of clean, oxygen-rich seawater. Favorable food conversion and rapid growth are other advantages of rearing *O. joubini*. On the negative side is the requirement of live food during their early stages. Although used in laboratories to study many aspects of animal behavior, learning, and morphology (vision and neurophysiology), their market is rather limited. Both squid and octopuses are popular research animals with scientists, as they are considered to be the most intelligent of the invertebrates.

REGIONS WHERE FARMED AND/OR RESEARCHED: University of Texas, Medical Branch, Galveston.

REFERENCES

Hanlon, R. T., and J. W. Forsythe. 1985. Advances in laboratory culture of octopuses for biomedical research. *Laboratory Animal Science* 35(1):33–40.

Hanlon, R. T., and R. F. Hixon. 1983. Laboratory maintenance and culture of octopuses and loliginid squids. In *Culture of marine invertebrates; selected readings*, ed. C. J. Berg, pp. 44–61. Stroudsburg, PA: Hutchinson Ross.

Squid

SCIENTIFIC AND COMMON NAMES: ***Loligo* spp., Squid**

DESCRIPTION AND DISTINCTIVE CHARACTERISTICS: Squid are close relatives of the octopus, and their bodies are elongate and torpedolike. They possess eight long arms plus two longer tentacles and lateral fins. They can change color and color pattern rapidly and are jet-propelled, moving by means of a funnel that can expel water under high pressure.

HABITAT RANGE: Widely distributed in the world's oceans from shallow, inshore waters. *L. opalescens* occurs in the eastern Pacific in the California Current. *L. pealei* is found in the western Atlantic (5–50° N), including the Gulf of Mexico and the Caribbean, near the surface down to about 1,200 ft.

REPRODUCTION AND DEVELOPMENT: Spawning occurs at different times in different locations: December–March off southern California, April–November off central California. Females may produce viable eggs at seven months of age. Development is direct: Young are tiny versions of adults.

AGE AND GROWTH: Male *L. opalescens* may reach about 7 in. and 4.5 oz; females about 6 in. and 3 oz. Males grow faster than females. Squid hatched in early summer become adults in roughly a year.

FOOD AND FEEDING: Eat shrimps, crabs, fishes, and other cephalopods.

PARASITES AND DISEASE: A mesozoan (dicyemid species) has been found in the eyes, stomach, digestive caeca, mantle cavity, and mesenteries of wild squid; infection rates of this parasite may reach over 75 percent. The lack of strontium in rearing water may cause squid to lose nervous control and spin. Bacteria can cause mortality in squid cultures.

PREDATORS AND COMPETITORS: The squid is an important prey for fish (>22 spp.), sharks, birds (13 spp.) and marine mammals (seals and small whales).

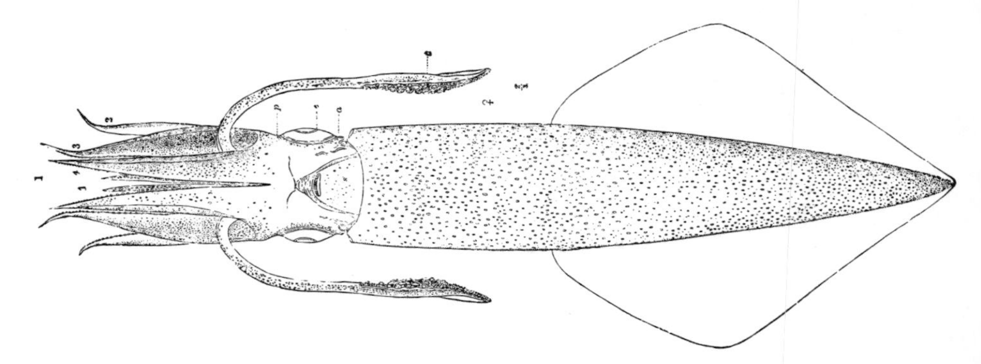

Squid species *Loligo pealei.*

AQUACULTURAL POTENTIAL: Squid are raised for their giant nerve fibers, which are widely used in medical research; they are also recommended for behavioral and genetic studies. *Loligo* has attributes of interest to aquaculturists: In general the large-egged squid are hardy and easy to spawn and raise in the laboratory. Favorable food conversion and rapid growth are other advantages of rearing squid. On the negative side, squid require live crustaceans to trigger feeding during their early stages.

REGIONS WHERE FARMED AND/OR RESEARCHED: University of Texas, Marine Biomedical Institute, Medical Branch, Galveston.

REFERENCES

Gilbert, D. L., W. J. Adelman, Jr., and J. M. Arnold, eds. 1990. *Squid as experimental animals.* New York: Plenum Press.

Hanlon, R. T. 1987. Mariculture. In *Cephalopod life cycles.* Vol. II: *Comparative reviews,* ed. P. L. Boyle, pp. 291–305. New York: Academic Press.

Hanlon, R. T., and R. F. Hixon. 1983. Laboratory maintenance and culture of octopuses and loliginid squids. In *Culture of marine invertebrates; selected readings,* ed. C. J. Berg, pp. 44–61. Stroudsburg, PA: Hutchinson Ross.

ORNAMENTAL SPECIES (PETS)

The ornamental fish industry in the United States is big business. In 1990 alone almost $41 million worth of fish were imported and over $11 million exported. Continued interest in home aquaria, advanced technology, and environmental concerns over collecting wild species for the industry could greatly strengthen the industry in some states, especially Florida. Pet stores handle large numbers of freshwater fish, 75 percent of which are farm-raised. The more colorful marine fish, which challenge the aquarists, also continue to gain in popularity. The value of the industry is greatly enhanced by the additional supplies and equipment purchased by aquarists in support of their hobby.

Plants

Interest in colorful plants for indoor aquaria and outdoor ponds has increased greatly. In Florida over 180 businesses carry aquatic plants for use in aquaria, water gardens (outdoor, freshwater ponds), and so on.

Plants make indoor aquaria more attractive and, with proper lighting, supply oxygen to fish, remove carbon dioxide, and perhaps even make the fish feel more comfortable. (In darkness plants remove oxygen needed by the fish.) The size of the aquarium dictates the size of the plants used: A few small plants are usually all that can be added to a small aquarium for a good balance. A number of different plant species are supplied to pet stores by wholesalers.

Water gardens have enjoyed a strong increase in interest. Florida alone has 33 nurseries growing suitable aquatic plants, with annual sales estimated at $7–8 million. One of the most common showy plants used in water gardens is the water lily, a common name applied to perennial herbs in the genus *Nymphaea*.

Water Hyacinth

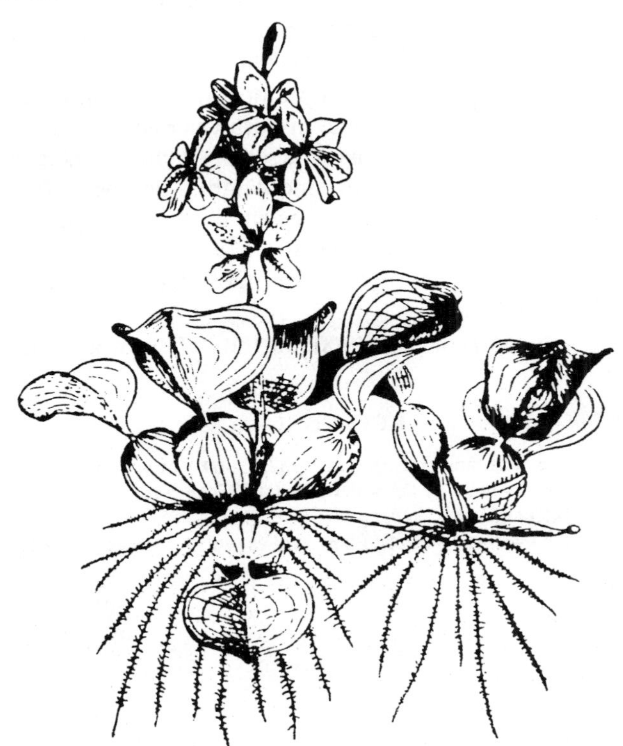

SCIENTIFIC AND COMMON NAMES: ***Eichhornia crassipes*, Water hyacinth**

DESCRIPTION AND DISTINCTIVE CHARACTERISTICS: A freshwater floating plant with adventitious roots.

HABITAT RANGE: Widely distributed from South America as cattle fodder: on the U.S. coast of the Gulf of Mexico; on the Atlantic Coast as far north as North Carolina; on the coast of central California; and in Hawaii and Bermuda.

REPRODUCTION AND DEVELOPMENT: Reproduction is by sexual and asexual seeds.

AGE AND GROWTH: Approximately 15 percent weight increase may be achieved in 7–13 days under favorable conditions.

FOOD AND FEEDING: These plants extract nutrients from water and, using sunlight, grow by photosynthesis.

PARASITES AND DISEASE: Fungi cause disease.

PREDATORS AND COMPETITORS: Snails and arthropods (insects) are the principal predators. Chinese grass carp, turtles, and manatees graze on water hyacinth leaves. Intraspecific competition results from high density; interspecific competition from loss of available nutrients to other water plants.

AQUACULTURAL POTENTIAL: Water hyacinths remove pollutants from fish-rearing facilities, and thus are valuable to freshwater aquaculturists. They can absorb relatively high concentrations of arsenic, cadmium, mercury, and lead rather rapidly, thus preventing high fish mortalities. Their beautiful flowers add color to fish ponds and tanks. Some fishes eat them under conditions of confinement. Note that it is illegal in some states (e.g., Florida) to possess water hyacinths because, given suitable conditions, they can reproduce rapidly, clogging lakes and waterways.

REGIONS WHERE FARMED AND/OR RESEARCHED: Mississippi; India.

REFERENCES

Chigbo, F. E., R. W. Smith, and F. L. Stone. 1982. Uptake of arsenic, cadmium, lead and mercury from polluted waters by the water hyacinth *Eichhornia crassipes*. *Environmental Pollution, Ser. A*, 27(1):31–6.

Gopal, B. 1987. *Water hyacinth*. Aquatic Plant Studies, vol. 1. Amsterdam: Elsevier.

Invertebrates

Arthropods/Crustaceans

Crabs

SCIENTIFIC AND COMMON NAMES: ***Coenobita clypeatus,*** Called *soldaatje* (little soldier) in the Netherlands Antilles. **Land hermit crab**

DESCRIPTION AND DISTINCTIVE CHARACTERISTICS: The *cephalothorax* (hard part of the body) bears the head, thoracic limbs, eyes, antennae, jaws, and gills, as well as an exoskeleton that protect it. Body color ranges from beige to orange-brown, with blue- or purple-tipped claws, the left claw being much larger than the right. Because the soft, vulnerable abdomen of the crab has no exoskeleton, the crab must find an empty shell to use as protection from its enemies, and to help keep itself moist.

HABITAT RANGE: From Bermuda southward; common in the Leeward and Windward Islands, Netherlands Antilles, southern Florida, and the Bahamas; also found in South America, from Venezuela to Brazil.

REPRODUCTION AND DEVELOPMENT: Eggs hatch and develop into a larval stage lasting about one month; thereafter they move onto land, returning to the sea only to deposit their eggs. A large female may have up to 5,000 eggs.

AGE AND GROWTH: May weigh over 1 lb (including shell). One land hermit lived 11 years in captivity.

FOOD AND FEEDING: Plant and animal matter, chiefly leaves and fruits, possibly dead fish; they may be cannibalistic.

PARASITES AND DISEASE: None reported.

PREDATORS AND COMPETITORS: Young are vulnerable to fish; birds, rats, and raccoons eat large juveniles and adults despite the sometime heavy shell they have appropriated. Mites are commensals inside the shell.

AQUACULTURAL POTENTIAL: In recent years land hermit crabs have been collected and kept as pets. These crabs are easy to grow and maintain, and their ability to climb on plants or twigs placed in the aquarium makes them attractive as pets. They require increasingly larger shells to live in as they grow larger, and their behavior as they inspect and move into a new shell is interesting to observe. They will drive another crab out of a shell that may appear to be better than the one they are in. Collectors have turned commercial—some on a small scale but others on a large scale, with outlets in major department stores and pet shops throughout the United States; however, the market is dependent on fads.

REGIONS WHERE FARMED AND/OR RESEARCHED: Southern Florida.

REFERENCES

De Wilde, P. A. W. J. 1973. On the ecology of *Coenobita clypeatus* in Curacao: with reference to reproduction, water economy and osmoregulation in terrestrial hermit crabs. *Studies on the Fauna of Curacao and Other Caribbean Islands* 44 (no. 144):1–138.

Iversen, E. S., and R. H. Skinner. 1977. *Land hermit crabs in nature and as pets*. Miami, FL: Windward Publishing.

Live Rocks

While the gathering from nature and sale of *live rocks* (rocks containing attached and encrusted organisms) to our knowledge is rather limited, it is nonetheless an interesting new industry that rode in on the shirttails of the increased popularity of marine aquaria. The first hobby-sized marine aquaria usually held only marine fish because attempts to keep other organisms alive, especially invertebrates, were unsuccessful. Instead, fish shared early home marine aquaria with a variety of decorative ornaments (mermaids, castles, caves, rocks, etc., made of painted ceramics). Improved equipment and increased knowledge of the requirements of colorful, fascinating invertebrates permitted them to join fish in aquaria, and they quickly became popular.

Live rocks support a variety of algae and invertebrates growing on their surface, and bacteria and boring organisms that permeate their mass. Important surface organisms are anemones, sponges, worms, gorgonians, and algae. Live rocks are found in the United States in nearshore tropical and subtropical waters, in the Gulf of Mexico from Mexico around the northern Gulf to the Florida Keys. On the southeastern U.S. coast, live rocks can be found as far north as southern Georgia. Southern California also has regions with live rocks.

Live Rock Collection

The live rock industry for years relied on collections from natural nearshore marine waters, primarily hard bottoms and a variety of reef forms. Because of potential severe damage to natural habitats associated with collecting live rocks, conservation agencies have recommended development of live rock aquaculture to reduce or remove the threat of environmental damage while satisfying the commercial demand for this product.

Types of live rock sought from specific habitats are rubble rock, algae rock, false coral, and sea mat. Once outside growth is removed by brushing, the rock is placed in an aquarium at proper temperature with an adequate wet/dry filtering system; it then "blooms" from within its pores. This technique prevents the death and decay of many encrusting organisms that frequently occurs in the change from a natural habitat to an aquarium and contaminates the aquarium water. "Cured" or "seeded" live rocks are those that have been collected in the wild and meet certain criteria; for instance, they must be coral rock or limestone of certain size and free of materials toxic to other animals. These are more expensive than fresh rocks ($4–6 vs. $3/lb), but may avoid many water-quality headaches for aquarists. Prices of live rock can be as low as 90 cents to about $20 per pound. They are shipped by air throughout the United States, and to Canada and England.

A debate has been going on for many years between the live rock collectors and state natural resource agencies concerning the taking and sale of live rocks. The position of the agencies is that the removal of commercial quantities of live rock from the state's submerged lands conflicts with the state's overall goal of protecting its natural resources. About 160,000 lb of decorative rocks were taken from the Florida Keys in 1988. Furthermore, collectors may destroy parts of reefs, take rock supporting endangered reef species, disrupt the food web (many of the invertebrates on live rocks serve as intermediate links in between primary producers and higher forms such as fishes), and endanger coral by sedimentation.

Live rock collectors claim that they remove only a tiny fraction of the rock present in state waters, and their business enables aquarium enthusiasts to enjoy a miniature reef ecology system in their homes. Over 50 permits were issued to remove live rock in the Florida Keys; another 20–30 persons are said to be active in the harvesting of live rocks, and six major shippers have been identified. Then, in 1989, the Florida Department of Natural Resources banned the removal of live rocks from the wild.

Live Rock Culture

Accepting that the traditional harvesting practices of live rocks is harmful to the environment, and could become much more so with increase in the industry, attention has turned to possible culture techniques. In field experiments, recruitment of sessile benthic organisms on natural bottom cleared by humankind or its actions (shipwrecks, etc.) has been very slow: Some types of substrate have to be conditioned for periods of up to a year before colonization occurs. In general, the condition and position of the substrate affects its colonization. For example, sand that tends to shift continuously can smother juvenile invertebrates. In some studies, there were no signs of significant colonization for some species even after 10 years. In summary, it seemed that to expect barren rocks in a natural marine environment to become colonized and ready for sale in a practical time is unreasonable, due to the different kinds of natural rocks and the wide variety of species with divergent life histories. However, in recent Florida Sea Grant–funded studies in Tampa Bay, investigators placed barren rocks on platforms that were *strategically located:* Areas with minimal currents and waves were chosen, so that sedimentation would not harm species during settlement and early growth on the rocks. These preliminary trials lasted only six months but suggested to the investigators that live rock culture could be economically and environmentally sound. This revised methodology holds promise for future private ventures.

To go one step further and try to produce commercial quantities of live rocks for profit in closed culture systems is even more difficult than using the open culture system mentioned above. Organisms that might loosely be called "brood stock" would have to be collected in the wild and maintained in tanks or ponds so that they can inoculate barren rocks for subsequent growth. Without adequate knowledge of the life histories of colonizing organisms—which is where we stand now—the outlook for successful closed culture commercial production of live rock is not encouraging.

REFERENCES

Blackburn W. 1988. The ecological use of live rock. *Freshwater and Marine Aquarium* 11(8): 8–11.

Delbeek, J. C. 1990. Live rock algal succession in a reef system. *Freshwater and Marine Aquarium* 13(10):120–35, 179.

Derr, M. 1992. Raiders of the reef. *Audubon* 94(2):48–56.

Feddern, H. A. 1990. Live rock for the marine aquarium; recruitment, erosion, harvest, and discussion. *Freshwater and Marine Aquarium* 13(12):102–6.

Nilsen, A.J. 1990. The successful coral reef aquarium. Part 4. *Freshwater and Marine Aquarium* 13(12):98–101.

Mollusks/Gastropods

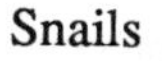

Snails

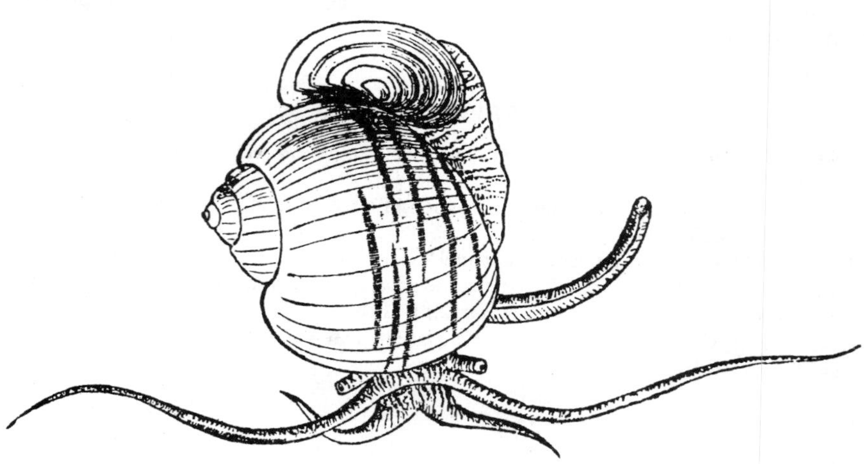

SCIENTIFIC AND COMMON NAMES: ***Pomacea (Ampullaria)* spp., Apple snail**
Related: *Pila* spp. (imported from China).
As many as 11 families of freshwater snails are found in North America.

DESCRIPTION AND DISTINCTIVE CHARACTERISTICS: Horn-shaped shell is banded; color varies between olive-green and yellow-brown.

HABITAT RANGE: Spend most of their lives in fresh water, but occasionally venture out on land (they possess gills and lungs); shallow, warm waters, such as ponds, rice fields, and swamps, seem to be preferred. Apple snails have been observed in temperatures over 100 °F. One species is found in tropical South America and the Caribbean, another in Africa, Madagascar, and Southeast Asia. Because they are herbivores, transplanting snails outside of their natural geographic range may introduce a species that will become a serious garden and farm pest. Federal government permits must be obtained for any anticipated transplants within the United States, or for imports of exotic snails.

REPRODUCTION AND DEVELOPMENT: Snails are hermaphrodites. Eggs are deposited on land or in aquaria just above the water line. Hatching takes place in about two weeks.

AGE AND GROWTH: Some snail species may live five years or more and reach a diameter of about 3 in.

FOOD AND FEEDING: Snails eat vegetable matter, but occasionally will eat decaying flesh. They possess a structure called a *radula* that is used as a rasp to eat attached algae. In captivity they will eat lettuce.

PARASITES AND DISEASE: Bacteria, nematodes, and trematodes.

PREDATORS AND COMPETITORS: Snails are eaten by many predators, including fishes, water birds, rats, mice, toads, lizards, and insects.

AQUACULTURAL POTENTIAL: Apple snails are sold to the aquarium trade not only as decorations, but also to remove the growth of algae that accumulates on the sides of fish aquaria: Usually a single snail in a fish aquarium will suffice to prevent annoying algal growth. However, perhaps because of their voracious appetite for aquatic plants and high metabolic rate, several apple snails together in an aquarium may cloud the water, making it unfit for fish. Young fish benefit from the presence of snails since they eat the *infusoria* (ciliated protozoans) that feed on the snail droppings. Snails serve as food for pond-reared centrarchid fishes (e.g., yellow perch, bluegills, pumpkinseeds); generally, no effort is made to increase the production of snails in centrarchid fish ponds. U.S. snail farms are small-scale operations catering to the aquarium trade.

Some species in this genus grow large enough and fast enough to be eaten by humans. French chefs prefer the freshwater snail (*Helix aspersa*) to prepare *escargots,* as they call food snails; they are very popular not only in France, but also in Japan. Marine snails, such as carnivorous moon snails and whelks, are also sometimes used in escargot recipes; however, these snails may pick up toxins when they feed on bivalves in areas where dinoflagellate blooms have occurred. This can cause paralytic shellfish poisoning (PSP) in humans who they eat the marine snails.

REGIONS WHERE FARMED AND/OR RESEARCHED: Snail Club of America, Fresno, California; Hawaii, Utah, Florida, Indiana, Arizona, Texas.

REFERENCES

Baratou, J. 1988. *Raising snails for food.* Translated from the French by Frances Herb. Callistoga, CA: Illuminations Press.

Cheney, S. 1988. *Raising snails.* Special Reference Briefs NAL SRB 88–04. Beltsville, MD: National Agricultural Library.

Socolof, R. B. 1980. Snail update..., and a new pale snail. *Freshwater and Marine Aquarium* 3(10):33–4.

Starmuhlner, F. 1989. The alluring apple snails. *Tropical Fish Hobbyist* 37(5):52–7.

Vertebrates

Ornamental Fish

SCIENTIFIC AND COMMON NAMES: Between 500 and 1,000 species of freshwater and marine finfish are sold in the aquarium business. These include tetras, Siamese fighting fish, swordtails, guppies, mollies, discus, clownfish, goldfish, koi, tilapias, and many others.

DESCRIPTION AND DISTINCTIVE CHARACTERISTICS: Small colorful fishes, usually with interesting behavioral patterns, such as stunning insects with water (archerfish), mouth brooding (some tilapias), kissing (gouramis), or nursing their young (discus). A variety of carp called *koi* has been expressly bred for special colors, including silver, gold, black, and white color patterns.

HABITAT RANGE: Mostly tropical and subtropical in fresh and salt water.

REPRODUCTION AND DEVELOPMENT: Ornamental fish may be nest builders, live bearers, or egg layers.

AGE AND GROWTH: Generally grow rapidly, from egg to adult within six to eight months. Koi are remarkably long-lived: The oldest on record in Japan lived 226 years.

FOOD AND FEEDING: Feed on a variety of foods in nature; can be herbivorous, carnivorous, or omnivorous. In captivity young are fed *Artemia* (brine shrimp) larvae; adults are fed commercially available dried and frozen foods.

PARASITES AND DISEASE: Parasite species vary among the many different aquarium species raised, depending on such factors as whether they derive from salt or fresh water, and were captured in the wild or bred from captive parents. Parasites range from protozoans (e.g., "Ich" or white spot disease) through virtually all groups to crustaceans (e.g., fish lice). Microorganisms (fungi, viruses, and bacteria) are responsible for serious diseases. Abiotic diseases are induced by poor nutrition, poor water quality, and crowding in small tanks.

PREDATORS AND COMPETITORS: In nature, larger fish are predators. Interspecific competition occurs in small aquaria.

AQUACULTURAL POTENTIAL: About 20 percent of the aquarium fishes in the United States are imported; the remainder are raised domestically. Florida is the only U.S. state where climate and water supplies are suitable to support outdoor tropical aquarium fishes; however, even in Florida occasional fish kills occur due to cold spells. In 1973, retail sales of ornamental fishes in the United States topped $200 million, and have since been increasing. On a value-per-pound basis, ornamental fish are way above all other cultured species; some sell for several hundred dollars a pound.

Certain colors and color patterns desired by koi fanciers command prices in the $10,000 range, and record prices paid for a single fish are 10 times as high.

Koi (*Cyprinus carpio*).

Because of their long life and high value, Japanese koi owners frequently include their fish in their wills. Collecting koi has come to the United States, where in recent years it boasts an increasing following. Koi are also important in Indonesia, China, Europe, and Israel.

Knowledge and experience are vital to the proper management of water quality for ornamental fish species. Considerable labor is required to grade and provide vigorous, disease-free fish. Furthermore, transporting ornamentals over long distances requires very careful handling, and frequently meeting inconvenient schedules of air carriers. Due to the genetic plasticity of many species, new strains of fish desirable to the aquarium trade can be developed. The market is seasonal, with highest sales during fall and winter, generally peaking around Christmas.

REGIONS WHERE FARMED AND/OR RESEARCHED: Florida; Singapore, Hong Kong, Bangkok; South America.

REFERENCES

Axelrod, H. R. 1973. *Koi of the world: Japanese colored carp.* Neptune City, NJ: T. F. H. Publications.

Axelrod, H. R., W. E. Burgess, N. Pronek, and J. G. Walls. 1986. *Dr. Axelrod's atlas of freshwater aquarium fishes.* Neptune City, NJ: T. F. H. Publications.

Axelrod, H. R., and L. P. Schultz. 1983. *Handbook of tropical aquarium fishes*, rev. ed. Neptune City, NJ: T. F. H. Publications.

Burgess, W. E., H. R. Axelrod, and R. E. Hunziker. 1988. *Dr. Burgess's atlas of marine aquarium fishes.* Neptune City, NJ: T. F. H. Publications.

Spotte, S. 1979. *Fish and invertebrate culture; water management in closed systems*, 2nd ed. New York: Wiley.

Winfree, R. A. 1989. Tropical fish; their production and marketing in the United States. *World Aquaculture* 20(3):24–30.

Goldfish

SCIENTIFIC AND COMMON NAMES: ***Carassius auratus*, Goldfish**

DESCRIPTION AND DISTINCTIVE CHARACTERISTICS: Similar to carp but differing in the dentition and the lack of barbels. Olivaceous coloration and habits are similar to carp. Aquarium specimens are very colorful (red and golds), having been bred for color first by the Chinese and later by the Japanese and Koreans; if released into natural waters they usually revert to their original color. The range of coloration is extensive, from white to black. Fin shapes, bulging eyes, misshapen heads, and spherical bodies are a few of the characteristics selected for breeding.

HABITAT RANGE: Native to China and Japan; transplanted widely. Prefer waters that are quiet with weedy bottoms.

REPRODUCTION AND DEVELOPMENT: Spawning is usually initiated when the water temperature reaches and stays above 60 °F; for some varieties, spawning might not commence until the temperature exceeds 70 °F. Some 2,000–4,000 eggs are released during each intermittent spawning, which may occur once every five days to every 48 hr. The eggs are attached to twigs or Spanish moss that serve as nests, and hatch in 2–7 days.

AGE AND GROWTH: May reach a weight of several pounds. Fancy goldfish may reach market size after three months.

FOOD AND FEEDING: Fed a diet of about 40 percent protein.

PARASITES AND DISEASE: Diseases are caused by bacteria (*Vibrio*, *Aeromonas salmonicida*), protozoans (*Mitraspora cyprini*, *Trichodina*), monogenetic trematodes (*Dactylogyrus*, *Gyrodactylus*), and anchor worms (*Lernaea*).

PREDATORS AND COMPETITORS: Aquatic insects and *Cyclops* are predators. A variety of large zooplankton may compete with young goldfish for food.

AQUACULTURAL POTENTIAL: Early records show that goldfish farming began in the Orient about 970 A.D. They were brought to the United States in the late 1800s. Since artificial culture began, a wide variety of fancy strains have been developed (e.g., black moors, bubble eyes, veintails, lion heads). Goldfish are used as bait fish in states where this is legal; they are considered ideal trotline bait, but not active enough as live bait for sportfishing. *Feeder fish* are what aquaculturists call goldfish raised to feed carnivorous aquarium fishes.

REGIONS WHERE FARMED AND/OR RESEARCHED: U.S. Southeast; Japan, China, Australia.

REFERENCES

Giudice, J. J., D. L. Gray, and J. M. Martin. 1981. *Manual for bait fish culture in the South.* Little Rock: University of Arkansas Cooperative Extension Service.

Martin, M. 1983. Gold fish farming. Part I. *Aquaculture Magazine* 9(3):38–40.

Martin, M. 1983. Gold fish farming. Part II. *Aquaculture Magazine* 9(4):38–40.

Martin, M. 1983. Gold fish farming. Part III. *Aquaculture Magazine* 9(5):30–34.

Smith, H. M. 1924. Goldfish and their cultivation in America. *National Geographic* 46:375–400.

Reptiles

Terrapin

SCIENTIFIC AND COMMON NAMES: ***Trachemys scripta,*** (=*Pseudemys*) Also called slider. **Red-eared terrapin**

DESCRIPTION AND DISTINCTIVE CHARACTERISTICS: There is a large yellow patch behind the eye. Shells show markings on the forepart and circular markings on the *plastron* (underside).

HABITAT RANGE: Inhabit freshwater ponds, slow-moving rivers, bayous, and swamplands from southeast Texas north to Illinois. Habitat includes all of the state of Louisiana and parts of Mississippi, Arkansas, Tennessee, Missouri, and Illinois.

REPRODUCTION AND DEVELOPMENT: Nest during June and July in western Tennessee. As few as five and as many as 22 eggs may be laid.

AGE AND GROWTH: The length of the carapace may reach 8 in. for males and 12 in. for females.

FOOD AND FEEDING: Terrapins eat small fish, insects, amphibians, snails, pond weeds, and carrion. They tend to eat more vegetation the larger they get.

PARASITES AND DISEASE: In captivity, blindness results from improper food, low temperature, and lack of sunlight.

PREDATORS AND COMPETITORS: Young are eaten by birds, large fish, mammals, raccoons, and alligators. Some mammals destroy turtle nests and eat the eggs.

AQUACULTURAL POTENTIAL: Starting in about 1957 thousands of these terrapins were raised and shipped out of Louisiana. To increase sales, their shells were painted and names or messages added. At one time as many as 150 farms were in operation. In 1975, a ban was placed on U.S. retail sales and importation of these turtles because of human-contracted *Salmonella* (bacterial) infection; a similar ban was imposed in Canada. Food given the turtles had contained *Salmonella,* a bacterium that causes fever, diarrhea, and vomiting in humans (the species *S. typhosa* causes typhoid fever). Some farms are now raising and packaging "*Salmonella*-free turtles" and seeking to have the ban lifted. Anyone interested in turtle farming should check local laws governing this industry. Canada's Department of Agriculture requires permits for live-turtle importation. Biological supply companies provide a small market for farmed turtles.

REGIONS WHERE FARMED AND/OR RESEARCHED: Louisiana, Texas, and Mississippi.

REFERENCE

Warwick, C. 1986. Red-eared terrapin farms and conservation. *Oryx* 20(4):237–40.

PRODUCTION OF CHEMICALS (CARRAGEENAN & DRUGS)

This section focuses on algae cultured for the extraction of chemicals, some of which may have application in human food. Macroalgae used directly as human food are discussed in Chapter 3.

Macroalgae

Macroalgae are the large plants growing in aquatic environments, whether salt, fresh, or brackish water. Many species grow along the sea coasts, mostly in the intertidal zones. The term *seaweed* is frequently used for the large marine algae, many of which are called *kelp*. These plants are classified by botanists according to several characteristics, including color: They may be blue-green, brown, or red, the latter two marine algae being the most valuable commercially.

Commercial algae, with few exceptions, are gathered in public areas. They are used as thickening agents in many processed foods, such as soups and ice cream, and are nutritious foods in their own right. Important algal products include agar, algin, and carrageenan. *Agar,* a gelling agent derived from red algae, is useful in the home and in the laboratory, especially as a culture medium. *Algin* (from brown algae), used in ice cream and chocolate milk drinks, has additional uses in manufacturing and in laboratories. Carrageenan (from red algae, especially Irish moss) is an emulsifier that finds use both in food (e.g., dairy products, puddings) and nonfood items (e.g., toothpaste, beer, pharmaceuticals, cough syrups, dental impression material, shampoos), as well as in industry (e.g., paper, paint, and rubber). The U.S. commercial algae industry is valued at $30 million. Up to seven times the U.S. production of some seaweeds are imported into this country to meet the demands of its carrageenan industry.

Seaweed cultivation has been most successful in Asiatic countries, which use inexpensive methods of seeding natural waters rather than the intensive raceway system used in North American experiments (and where the harvests are traditionally used for human food). Algae are good additions to some farming facilities and can fit nicely into a polyculture program. The production of methane as an additional source of energy has sparked macroalgal experimentation.

REFERENCES

Bird, K. T., and P. H. Benson, eds. 1987. *Seaweed cultivation for renewable resources.* Developments in Aquaculture and Fisheries Science, vol. 16. Amsterdam: Elsevier.

Mathieson, A. C. 1982. Seaweed cultivation: A review. In *Proceedings of the Sixth U.S.–Japan Meeting on Aquaculture, Santa Barbara, California, August 27–28, 1977,* ed. C. J. Sindermann, pp. 25–66. NOAA technical report NMFS-CIRC 442. Seattle: National Marine Fisheries Service.

Mitsui, A., and C. C. Black, eds. 1982. *CRC handbook of biosolar resources,* vol. 1, pts. 1 & 2, "Basic principles." Boca Raton, FL: CRC Press.

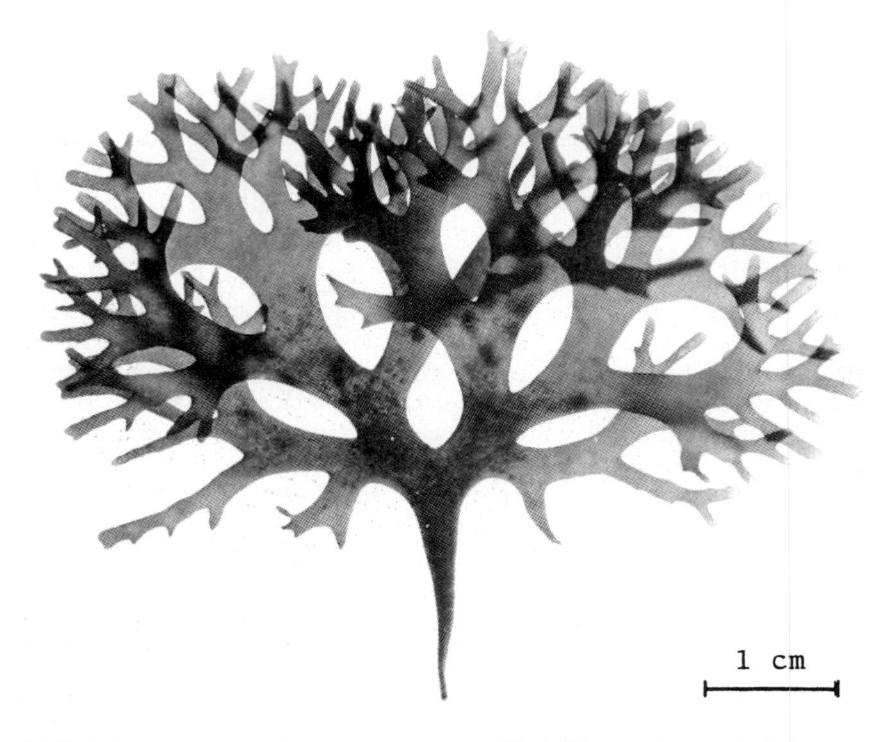

SCIENTIFIC AND COMMON NAMES: ***Chondrus crispus,* Irish moss**

DESCRIPTION AND DISTINCTIVE CHARACTERISTICS: A small, perennial marine red alga about 10 in. high that forms a thick carpet over rocks and ledges.

HABITAT RANGE: Atlantic Coast of North America, from Nova Scotia to Rhode Island, and the coasts of Scotland and Ireland.

REPRODUCTION AND DEVELOPMENT: Life history is carried out in culture.

AGE AND GROWTH: Growth is slow—only a tiny fraction of an inch per day during the summer; growth in suitable conditions (temperatures of 41–59 °F) is considerably faster than in the wild. Weight increases as high as sixfold per month have been achieved with low-density trials. Growth results from increase in the number of plants, rather than in the production of large plants.

FOOD AND FEEDING: Irish moss grows by photosynthesis, extracting nutrients and minerals from shallow water. Plants can absorb enough nitrogen in a few hours to carry them for several days. The cost of fertilizing the algal growing ponds can be reduced by stopping water exchange when nutrients are introduced into the pond.

PARASITES AND DISEASE: Bacteria, fungi, and nematodes are known to cause disease in Irish moss.

PREDATORS AND COMPETITORS: Sea urchins are predators in the wild; in culture, snails, isopods, and amphipods. Epiphytes can be a major problem, as they compete for nutrients and light in the growing ponds.

AQUACULTURAL POTENTIAL: Extract from Irish moss (carrageenan) is widely used in food processing and other industries. Harvesting is done either by raking attached plants loose from underwater beds or by simply collecting plants washed up on the beach by heavy weather. Special strains of Irish moss will grow detached (they lack holdfasts). Labor demands are generally low, but cost of culturing can be high for aeration and fertilizers used to stimulate growth in enclosures. Experimental culturing has been done in Canada, but additional research is needed before large-scale culture can be economical.

REGIONS WHERE FARMED AND/OR RESEARCHED: New York, New Hampshire; Canada (Nova Scotia).

REFERENCES

Bird, K. T., and P. H. Benson, eds. 1987. *Seaweed cultivation for renewable resources.* Developments in Aquaculture and Fisheries Science, vol. 16. Amsterdam: Elsevier.

Harvey, M. J., and J. McLachlan, eds. 1973. *Chondrus crispus.* Halifax: Nova Scotian Institute of Science.

Isaacs, F. 1990. Irish moss aquaculture moves from lab to marketplace. *World Aquaculture* 21(2):95–7.

Mathieson, A. C. 1982. Seaweed cultivation: A review. In *Proceedings of the Sixth U.S.–Japan Meeting on Aquaculture, Santa Barbara, California, August 27–28, 1977,* ed. C. J. Sindermann, pp. 25–66. NOAA Technical Report NMFS-CIRC 442. Seattle: National Marine Fisheries Service.

Yarish, C., C. A. Penniman, and P. van Patten, eds. 1990. *Economically important plants of the Atlantic: Their biology and cultivation.* Groton, CT: Connecticut Sea Grant College Program.

SCIENTIFIC AND COMMON NAMES: ***Macrocystis pyrifera*, Giant kelp**
Related: *Laminaria* spp.

DESCRIPTION AND DISTINCTIVE CHARACTERISTICS: A huge brown alga that may attain a height of 100–150 ft.

HABITAT RANGE: Central Baja, California, to Alaska in cold nearshore waters, down to 150 ft. The large fronds have flotation devices to keep them near the surface of the water, enabling them to photosynthesize.

REPRODUCTION AND DEVELOPMENT: Giant kelp have special reproductive leaves that produce biflagellated zoospores; these are released into the water, where they swim about and settle to the bottom. The spores grow into tiny plants that produce eggs and sperm, which unite to grow into the adult plants.

AGE AND GROWTH: A full-grown plant may be six years old.

FOOD AND FEEDING: Nutrients are extracted from the water for photosynthesis.

DISEASES AND PARASITES: Bacteria cause disease.

PREDATORS AND COMPETITORS: Sea otter and sea urchin (*Strongylocentrotus franciscanus*) are predators, as are other echinoids, gastropods, and crustaceans.

AQUACULTURAL POTENTIAL: Giant kelp grow rather slowly. The fronds near the surface of the sea can be harvested and will regenerate, so continuous production can be realized. The plants are not eaten directly but yield several commodities: additives for human food, fertilizers, stabilizers in plants and drinks, and alginic acid (= algin). Years ago kelp was an important source of both potash and acetone.

REGIONS WHERE FARMED AND/OR RESEARCHED: Along the Pacific Coast of North America, especially California. A related *Macrocystis* species is farmed on Canada's western coast.

REFERENCES

Bird, K. T., and P. H. Benson, eds. 1987. *Seaweed cultivation for renewable resources.* Developments in Aquaculture and Fisheries Science, vol. 16. Amsterdam: Elsevier.

North, W. J., and M. Neushul. 1968. A note on the possibilities of large scale cultivation of *Macrocystis. Fish Bulletin* (California Dept. of Fish and Game) 139:17–24.

MISCELLANEOUS MARKET PRODUCTS

Pearls

Pearls are formed by certain species of freshwater and marine bivalve mollusks (e.g., oysters, freshwater mussels, and pen shells, and a few gastropod species of abalone and conch), and are an abnormal growth resulting from the invasion of the body of the mollusk by foreign matter, such as a particle of sand, which is considered to be an irritant to the animal. The animal coats the foreign matter with layers of *nacre,* the hard, iridescent that forms the inner layer of a molluscan shell (also called *mother-of-pearl*). Pearl values are determined by their luster, size, shape, smoothness, and colors.

The Chinese are credited with discovering the techniques of culturing pearls during the fourteenth century, but it was Kokichi Mikimoto in Japan who, in 1883, pioneered cultured pearls and gained a worldwide reputation for his product. Mikimoto's 1919 patented process requires great skill and patience. A small piece of the mantle of the mollusk, along with a small spherical shell fragment that will be the nucleus of the pearl, is placed in the oyster. The piece of *mantle* (graft tissue) grows around the nucleus to form a pearl sac. This may take several years, and as few as 5 percent of the surviving oysters may have pearls suitable for market. During the dawn of pearl culture the nuclei tried were mud pellets, nacre beads, and even small leaden images of the Buddha. Mikimoto found the best nuclei, those with the right material density and a desirable color white as the pearls themselves, were two species of freshwater mussels from the Mississippi River. The shells are obtained from wild mussels, cleaned, and shipped to Japan, where they are processed into spherical beads and later implanted in the marine pearl oyster.

Gemologists report that the "pearls" infrequently produced by queen conchs are not true pearls, as the calcareous concretions lack mother-of-pearl. Moreover, the odds of finding one are about 1 in 10,000. Nevertheless, some very valuable, good quality conch "pearls" have been found: One sold for just under $12,000 in 1984, and in 1987 another large one sold for $4,400. One disadvantage of conch "pearls" is their tendency to fade when exposed to sunlight.

During the 1930s two biologists working in Key West, Florida—La Place Bostwick and Clarence Hoy—claimed to have learned how to culture conch "pearls." There was no evidence, however, that this was so, or that the "pearls" they produced had actually been cultured. More recently, attempts to culture "pearls" have been underway at the Turks and Caicos conch hatchery and farm. One thousand conch have been seeded, and some preliminary observations suggest a degree of success may be achieved.

Like other mollusks, abalone will deposit nacre around a foreign substance when it becomes lodged between the abalone's shell and mantle. This may result in a "free pearl" (one not attached to the shell). These abalone pearls are usually irregularly shaped and of value only as novelties. In 1925 L. Bouton, a Frenchman, told of a method he developed as early as 1898 for producing pearls in abalone; his technique was adapted by the Japanese to their cultured pearl oyster industry. Bostwick, besides studying queen conch "pearl" culture, experimented with the culture of abalone pearls at the Scripps Institution in California during 1934–40; his results were never made public. A Dr. P. Fankboner of Simon Frazer University in British Columbia is reported to have developed techniques to produce abalone pearls, and abalone farmers in California and Hawaii have recently been offered the opportunity to have their abalone seeded for a fee and a share in the sale of gems produced.

Freshwater pearls from mussels are cultured in Tennessee, Texas, Louisiana, and California. Pearls are also raised in a land-based facility in Hawaii using the pearl oyster.

REFERENCES

Fassler, C. R. 1991. The return of the American pearl; three feisty farmers take on the Japanese. *Aquaculture Magazine* 17(6):63–78.

Fritsch, E., and E. B. Misiorowski. 1987. The history and gemology of queen conch "pearls." *Gems & Gemology* 23(4):208–21.

Fritsch, E., and E. B. Misiorowski. 1988. Queen conch "pearls": Rare gems from the Caribbean Sea. *Sea Frontiers* 33(5):286–91.

SCIENTIFIC AND COMMON NAMES: ***Pinctada martensii*, Pearl oyster**

DESCRIPTION AND DISTINCTIVE CHARACTERISTICS: About 4 in. when fully grown. Shells are convex and yellow-gray, with seven or so purplish-brown stripes.

HABITAT RANGE: Nearshore waters of Japan.

REPRODUCTION AND DEVELOPMENT: Spawning begins when the water temperature reaches about 77 °F—usually late June or early July—and normally ends by August. Sex reversal occurs in pearl oysters.

AGE AND GROWTH: Larval stages may take about 25 days. Postlarval growth is usually about six months, by which time pearl oysters reach about 1.25 in. They are harvested for pearls at age 3–4 years, reaching about 3.13 in. at age 4.

FOOD AND FEEDING: Filter feeders.

PARASITES AND DISEASE: Mortalities are caused by red tides, mostly dinoflagellates. Barnacles may attach near the hinge and interfere with opening and closing of the oyster, causing death.

PREDATORS AND COMPETITORS: Preyed upon by a number of fishes and shellfish, including the common eel (*Anguilla japonica*) and octopus. Seaweeds and sponges may be competitors for food and space and interfere with the ability of oysters to close their valves.

AQUACULTURAL POTENTIAL: *P. martensii* is the species used in the very successful Japanese pearl culture started in 1894 by Mikimoto. Of 100 oysters seeded, about 20 will produce marketable pearls; of those, about 5 or so will be gems, and of these, very few will be of exceptional value. The industry requires trained, careful workers who will perform routine, painstaking tasks day after day. Security is a major problem.

REGIONS WHERE FARMED AND/OR RESEARCHED: Japan, Papua New Guinea, Australia, Thailand, and Hawaii.

REFERENCES

Cahn, A. R. 1949. *Pearl culture in Japan*. Fishery Leaflet no. 357. Washington, DC: U.S. Fish and Wildlife Service.

Ward, F. 1985. The pearl. *National Geographic* 168:193–222.

Sponges

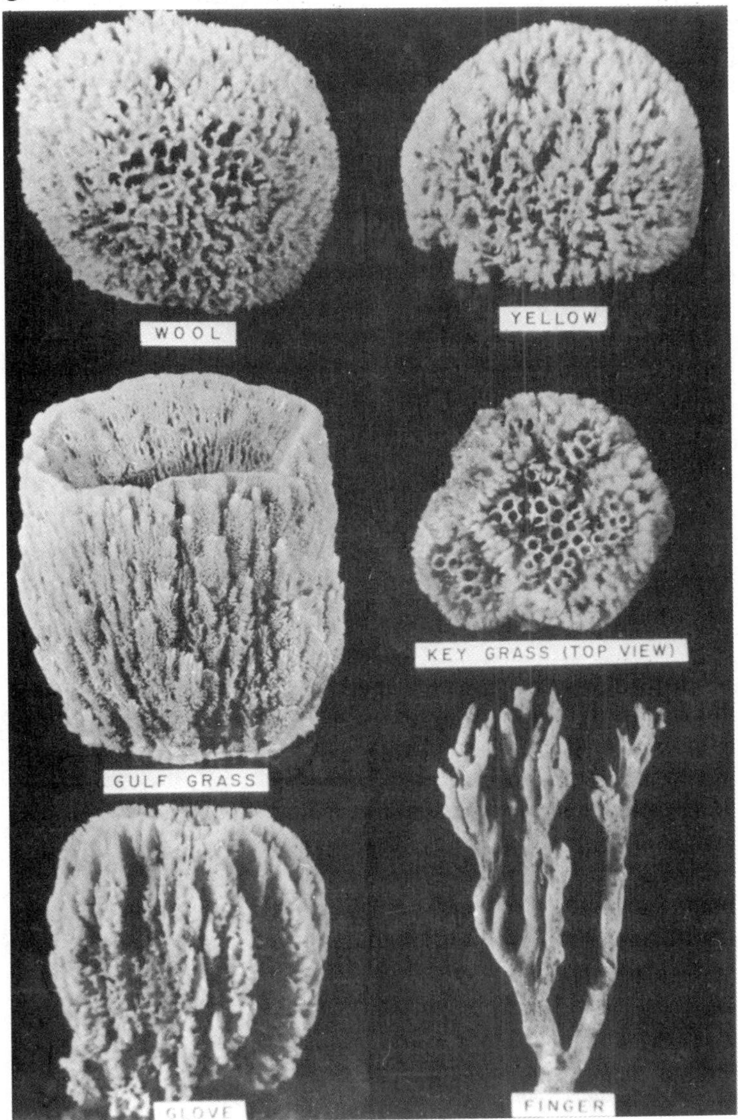

SCIENTIFIC AND COMMON NAMES: ***Hippiospongia lachne*, Wool sponge**
Related: yellow sponge (*Spongia barbara*), grass sponge (*S. graminea*); all belong to the phylum Porifera.

DESCRIPTION AND DISTINCTIVE CHARACTERISTICS: Heavy, slimy, sessile (permanently attached) animals, varying in color from light gray-yellow to brown or black. They are simple animals made up of a variety of specialized cells, and have many inhalant canals that carry oxygen, food, and water in, and exhalant canals that carry wastes out of the animal.

HABITAT RANGE: Inhabit warm marine waters (50–95 °F), some species in water as deep as 560 ft. Found on the western coast of Florida and in Florida Bay, Biscayne Bay, and the Bahamas.

REPRODUCTION AND DEVELOPMENT: Sponges reproduce by regeneration, budding, and sexual means. In sexual reproduction, a larval stage is produced that, after a short drifting period, must attach to a clean object.

AGE AND GROWTH: Sponges grow rather slowly; some four to seven years are required from a cutting to a market-sized sponge.

FOOD AND FEEDING: Detritus and plankton filter feeders.

PARASITES AND DISEASE: During 1938–9 there was an epizootic of commercial sponges in the West Indies and the Gulf of Mexico caused by a "funguslike" disease. In 1947–8 a disease of unknown origin affected commercial sponges on the western coast of Florida. They are also killed by red tide.

PREDATORS AND COMPETITORS: Sponge crabs burrow into living sponges, and many organisms live on or in them; these are considered shelter-seeking, not parasitic, but nevertheless reduce or destroy the commercial value of the sponges so inhabited.

AQUACULTURAL POTENTIAL: The slow growth of natural sponges, the high labor costs required for setting out cuttings, and the popularity of synthetic sponges make the prospects of sponge farming poor. A fisher who "catches" a sponge may gain more economically by selling it rather than making cuttings, setting them out, and waiting years for them to grow to market size.

REGIONS WHERE FARMED AND/OR RESEARCHED: As long ago as 1863 it was known that sponges could regenerate themselves when cut up. Experiments were carried out on the Island of Lesina by the Austrian government. Other unsuccessful attempts were made in the Florida Keys, British Honduras, and the Bahamas.

REFERENCE

Storr, J. F. 1964. *Ecology of the Gulf of Mexico commercial sponges and its relation to the fishery.* U.S. Fish and Wildlife Service Special Scientific Report no. 466. Washington, DC: U.S. Fish and Wildlife Service.

Selected References

BOOKS

Allen, P. G., L. W. Botsford, A. M. Schuur, and W. E. Johnston. 1984. *Bioeconomics of aquaculture*. Developments in Aquaculture and Fisheries Science, vol. 13, Amsterdam: Elsevier.

Aquaculture Magazine. 1991. *Aquaculture Magazine buyer's guide '92 and industry directory; 21st annual buyer's guide*, Asheville, NC: Aquaculture Magazine.

Arnold, C. R., G. J. Holt, and P. Thomas, eds. 1988. *Red drum aquaculture; proceedings of a symposium on the culture of red drum and other warm water fishes, Corpus Christi, TX, 1987*, Contributions in Marine Science, Supplement, vol. 30, Port Aransas, TX: University of Texas at Austin, Marine Science Institute.

Balchen, J. G., ed. 1987. *Automation and data processing in aquaculture*, IFAC proceedings series, no. 9, Oxford (U.K.): Pergamon Press.

Bird, K. T., and P. H. Benson. 1987. *Seaweed cultivation for renewable resources*, Developments in Aquaculture and Fisheries Science, vol. 16, Amsterdam: Elsevier.

Bourne, N., C. A. Hodgson, and J. N. Whyte. 1989. *A manual for scallop culture in British Columbia*, Canadian Technical Report of Fisheries and Aquatic Sciences, no. 1694, Nanaimo, BC: Fisheries and Oceans, Canada.

Brackishwater Aquaculture Information System. 1988. *Biology and culture of* Penaeus monodon, BRAIS state-of-the-art series, no. 2, Tigbauan, Iloilo (Philippines): Aquaculture Dept., Southeast Asian Fisheries Development Center.

Brown, E. E., and J. B. Gratzek. 1980. *Fish farming handbook*, Westport, CT: AVI Publishing Co.

Burgess, W. E., H. R. Axelrod, and R. E. Hunziker. 1988. *Dr. Burgess's atlas of marine aquarium fishes*, Neptune City, NJ: T. F. H. Publications.

Chamberlain, G.W., R. J. Miget, and M. G. Haby. 1990. *Red drum aquaculture*, Galveston: Texas Sea Grant College Program.

— eds. 1987. *Manual on red drum aquaculture*, Corpus Christi, TX: Texas Agricultural Extension Service, Texas A&M University.

Cobb, J. S., and B. F. Phillips, eds. *The biology and management of lobsters*. 2 vols. New York: Academic Press.

Conte, F. S., S. I. Doroshov, and P. B. Lutes. 1988. *Hatchery manual for the white sturgeon,* Acipenser transmontanus *Richardson, with application to other North American Acipenser-*

idae, Cooperative Extension University of California, Division of Agriculture and Natural Resources, pub. no. 3322, Oakland, CA: Division of Agriculture and Natural Resources, University of California.

Gordon, K., ed. 1989. *The North American directory of aquaculture, 1989–1990*, Vancouver, BC: Kevgor Aquasystems.

Hahn, K. O., ed. 1988. *Handbook of culture of abalone and other marine gastropods*, Boca Raton, FL: CRC Press.

Hoff, F. H., and T. W. Snell. 1989. *Plankton culture manual*, 2nd ed., Dade City, FL: Florida Aqua Farms.

Huguenin, J. E. 1989. *Design and operating guide for aquaculture seawater systems*, Developments in Aquaculture and Fisheries Science, vol. 20, Amsterdam: Elsevier.

Huner, J. V., and E. E. Brown, eds. 1985. *Crustacean and mollusk aquaculture in the United States*, Westport, CT: AVI Publishing Co.

Johnson, L., and B. Burns, eds. 1984. *Biology of the Arctic charr: Proceedings of the international symposium on Arctic charr*, Winnipeg: University of Manitoba Press.

Jung, C. K., and W. G. Co. 1988. *Prawn culture: Scientific and practical approach*, Dagupan City (Philippines): Westpoint Aquaculture Corp.

Kinne, O., ed. 1980–90. *Diseases of marine animals*, 4 vols., Hamburg (Germany): Biologische Anstalt Helgoland.

Laird, L. M., and T. Needham. 1988. *Salmon and trout farming*, New York: John Wiley.

Landau, M. 1991. *Introduction to aquaculture*. New York: John Wiley.

Lannan, J. E., R. O. Smitherman, and G. Tchobanoglous, eds. 1986. *Principles and practices of pond aquaculture*, Corvallis: Oregon State University Press.

Lovell, R. T. 1988. *Nutrition and feeding of fish*, New York: Van Nostrand Reinhold.

McLarney, W. O. 1984. *The freshwater aquaculture book: A handbook for small scale fish culture in North America*, Point Roberts, WA: Hartley & Marks.

McVey, J. P., ed. 1983. *CRC handbook of mariculture. Vol. I: Crustacean aquaculture*, Boca Raton, FL: CRC Press.

— ed. 1991. *CRC handbook of mariculture. Vol. II: Finfish aquaculture*, Boca Raton, FL: CRC Press.

Main, K. L., and W. Fulks. 1990. *The culture of cold-tolerant shrimp; proceedings of an Asian-U.S. workshop on shrimp culture, 1989*, Honolulu, HI: Oceanic Institute.

Manzi, J. J., and M. Castagna, eds. 1989. *Clam mariculture in North America*, Developments in Aquaculture and Fisheries Science, vol. 19, Amsterdam: Elsevier.

Meade, J. W. 1989. *Aquaculture management*, New York: Van Nostrand Reinhold.

Menzel, W., ed. 1991. *Estuarine and marine bivalve culture*, Boca Raton, FL: CRC Press.

Michaels, V. K. 1988. *Carp farming*, Farnham (U.K.): Fishing News Books.

Milne, P. H. 1978. *Fish and shellfish farming in coastal waters*, New York: State Mutual.

Nash, C.E., ed. 1991. *Production of aquatic animals: Crustaceans, molluscs, amphibians, and reptiles*. Amsterdam: Elsevier.

Pillay, T. V. R. 1990. *Principles and practice of aquaculture*, Farnham (U.K.): Fishing News Books.

Piper, R. G., I. B. McElwain, L. E. Orme, J. P. McCraren, L. G. Fowler, and J. R. Leonard. 1982. *Fish hatchery management*, Washington, DC: U.S. Fish and Wildlife Service.

Quayle, D. B., and G. F. Newkirk. 1989. *Farming bivalve molluscs: Methods for study and development*, Advances in World Aquaculture, vol. 1, Baton Rouge, LA: World Aquaculture Society.

Roberts, R. J., and C. J. Shepherd. 1986. *Handbook of trout and salmon diseases*, 2nd ed., Farnham (U.K.): Fishing News Books.

Sedgwick, S. D. 1988. *Salmon farming handbook*, Farnham (U.K.): Fishing News Books.

Shumway, S.E., ed. 1991. *Scallops: Biology, ecology and aquaculture,* Developments in Aquaculture and Fisheries Science, vol. 21. Amsterdam: Elsevier.

Sindermann, C. J. 1990. *Principal diseases of marine fish and shellfish,* San Diego: Academic Press.

Sindermann, C. J., and D. V. Lightner, eds. 1988. *Disease diagnosis and control in North American marine aquaculture,* 2nd rev. ed., Developments in Aquaculture and Fisheries Science, vol. 17, Amsterdam: Elsevier.

Spotte, S. 1979. *Fish and invertebrate culture: Water management in closed systems,* 2nd ed., New York: John Wiley.

Steffens, W. 1989. *Principles of fish nutrition,* New York: Halstead Press.

Stevenson, J. P. 1987. *Trout farming manual,* 2nd ed., Farnham (U.K.): Fishing News Books.

Stickney, R. R., ed. 1986. *Culture of nonsalmonid freshwater fishes,* Boca Raton, FL: CRC Press.

Tave, D. 1986. *Genetics for hatchery managers,* Westport, CT: AVI Publishing Co.

Texas Aquaculture Association. 1990. *Inland aquaculture handbook,* Austin, TX: Texas Aquaculture Association.

Tiddens, A. 1990. *Aquaculture in America: The role of science, government and the entrepreneur,* Boulder, CO: Westview Press.

Vakily, J. M. 1989. *The biology and culture of mussels of the genus* Perna, ICLARM Studies and Reviews, vol. 17, Manila (Philippines): International Center for Living Aquatic Resources Management.

Villalon, J. R. 1991. *Practical manual for semi-intensive commercial production of marine shrimp,* Galveston: Texas Sea Grant Program.

Webb, G. J. W., S. C. Canolis, and P. J. Whitehead, eds. 1987. *Wildlife management: Crocodiles and alligators,* Chipping Norton, N.S.W. (Australia): Surrey Beatty.

Wheaton, F. W. 1985. *Aquacultural engineering,* Malabar, FL: Krieger Publishing Co.

JOURNALS (Subscription prices are subject to change)

Aquacultural Engineering
Elsevier Science Publishing Co.,
Journal Information Center
655 Avenue of the Americas
New York, NY 10010
Bimonthly, $230.00/yr

A scholarly journal concerned with systems engineering for aquacultural producers. Covers facilities engineering and design, research-based studies, construction, operations, materials, and biological data.

Aquaculture
Elsevier Science Publishing Co.,
Journal Information Center
655 Avenue of the Americas
New York, NY 10010
$1182.00/yr for 8 vols. (32 issues)

A scholarly journal that covers worldwide research and management of aquaculture resources. Includes some short technical notes and book reviews.

Aquaculture and Fisheries Management
Blackwell Scientific Publishers, Inc.,
Journal Subscription Dept.
Marston Book Services
P.O. Box 87, Osney Mead
Oxford OX2 ODT, England
Quarterly, $277.50/yr

An international journal that publishes papers on theoretical and practical problems of marine and freshwater aquaculture, fish conservation, and the management of recreational and commercial fisheries worldwide.

Aquaculture Magazine
Subscription Dept.
P.O. Box 2329
Asheville, NC 28802
Bimonthly, $15.00/yr

A practical trade magazine covering the field in the U.S. and abroad. Includes advertisements for equipment and supplies, as well as classifieds, book reviews, and lists, reports of meetings, course announcements, and marketing statistics. An extensive annual *Buyer's guide* contains a directory of products and services, personnel, and addresses of state, federal, and regional aquaculture organizations.

Fish Farming International
Heighway Ltd.
33/39 Bowling Green Lane
London EC1R ODA, England
Monthly, $50.00/yr

A newspaper format with a European emphasis, though it surveys the global industry. Covers news of products and equipment, economic developments, research, conferences, and meetings. Publishes many advertisements, including classifieds.

Freshwater and Aquaculture Contents Tables
Fishery Information, Data & Statistics Svc.
Fisheries Dept., FAO
Via delle Terme di Caracalla
00100 Rome, Italy
Monthly, Free

Prints the tables of contents of core journals in freshwater science and aquaculture. Includes schedules of future conferences and meetings.

Journal of Applied Aquaculture
Haworth Press, Inc.
10 Alice St.
Binghamton, NY 13904-1580
Quarterly, $28.00/yr

This international journal, which began publishing in 1991, emphasizes applied research but includes developments in basic research as well. It encourages articles on new equipment and innovative techniques.

Journal of the World Aquaculture Society
World Aquaculture Society
143 J. M. Parker Coliseum
Louisiana State University
Baton Rouge, LA 70803
Quarterly, $75.00/yr (but included in Society membership)

A scholarly journal covering the scientific and technological development of worldwide aquaculture, especially culture systems, physiology, genetics/breeding, nutrition, disease, water quality, marketing, and economics.

Marine Science Contents Tables
Fishery Information, Data & Statistics Svc.
Fisheries Dept., FAO
Via delle Terme di Caracalla
00100 Rome, Italy
Monthly, Free

Reproduces the tables of contents of core journals in the marine sciences, including aquaculture. Also includes a schedule of conferences and meetings.

Northern Aquaculture
(formerly *Canadian Aquaculture*)
In Canada: Subscription Dept.
4611 William Head Road
Victoria, BC V8X 3W9
In U.S.: Harrison House Publishers
264 H Street, P.O. Box 8110-311
Blaine, WA 98230
7 issues/yr, U.S.$20.00, Can.$24.00

Focuses on northern species such as salmon, trout, mussels, and oysters. Includes news of research and technical developments from the U.S., Canada, and the world. Covers individuals and companies, equipment and supplies, and new species, and includes advertisements.

Progressive Fish Culturist
American Fisheries Society
5410 Grosvenor Lane, Suite 110
Bethesda, MD 20814
Quarterly, $19.00/yr

Practical as well as research articles on North American aquaculture. Includes short communications and technical notes as well as longer articles. Most authors are employed by universities, or by state or federal fish and wildlife departments.

Water Farming Journal
Carroll Trosclair & Associates
3400 Neyrey Drive
Metairie, LA 70002
Monthly, $19.00/yr

A useful trade journal with an emphasis on the U.S. industry. Includes advertising, classifieds, an international calendar of aquaculture events, and a book-ordering service.

World Aquaculture
World Aquaculture Society
143 J. M. Parker Coliseum
Louisiana State University
Baton Rouge, LA 70803
Quarterly, $30.00/yr (but included in Society membership)

Mostly technical reviews of the aquaculture industry and reports on significant trends in research and technology. Short notes and news items, book lists and reviews, a conference calendar, and advertisements are included.

World Shrimp Farming
(formerly *Aquaculture Digest*)
Aquaculture Digest
9434 Kearny Mesa Road
San Diego, CA 92131
Bimonthly, $60.00/yr

Prints short news releases covering individual farmers and companies, published articles of interest to the industry, notices of new books, practical tips, and economic trends. An annual report and directory are included in the subscription.

Appendix A

Glossary

Acanthocephala: Spiny-headed worms that, as adults, are intestinal parasites of vertebrates.

Adductor: Muscle that brings one body part toward another; e.g., a muscle that contracts to close the two halves of a bivalve's shell.

Adipose fin: Fatty tissue behind the dorsal fin of some fishes, including salmonids and catfishes.

Adult: Any animal that has attained full growth or is sexually mature.

Adventitious: Accidental; found in unusual places.

Aeration: Addition of oxygen or air to water; used in aquaculture to intensify production by maintaining good water quality.

Alga: Freshwater or marine chlorophyll-bearing plant (pl.: *algae*) ranging in size from a few microns to many feet in length; single-celled, colonial, or filamentous without true leaf, root, or stem systems.

Amnesic shellfish poisoning (ASP): Blue mussels in Prince Edward Island (Canada) poisoned 129 people and caused two deaths; symptoms included memory loss and disorientation; caused by the neurotoxin domoic acid, which is apparently produced by a diatom eaten by the mussels.

Amphipods: Small marine and freshwater crustaceans (including sandhoppers, sideswimmers, etc.) with compressed bodies, elongated abdomens, and no carapace.

Anadromous: Maturing and living most of one's life in the sea, but returning to fresh water to spawn (e.g., salmon).

Anchor worms: *See Lernaea.*

Anguillicola: Parasitic nematode of eels.

Annelids: Large group of segmented worms.

Aquaculture: Raising plants or animals in fresh, brackish, or salt water; also called mariculture (for seawater), fish ranching, fish farming.

Arthropods: Segmented animals with jointed legs and a chitinous exoskeleton; includes crustaceans, insects, horseshoe crabs.

Artificial reef: Man-made structure that attracts fish and is considered a form of aquaculture.

Ascidians: Order of tunicates; sea squirts.

Backyard aquaculture: Small-scale operation carried out by individuals or small communities for home food consumption.

Barbel: Slender, tactile structure near the mouth of certain fishes, such as catfishes.

Beko disease: Microsporean infection of fish muscles.

Bedding oysters: Small oysters, usually under about 50 mm (2 in.) long, suitable for rearing to maturity.
Benthic (benthonic) fauna: Organism living on or near the bottom of a body of water.
Bioassay: Quantitative measurement, under controlled conditions, of the effects of a substance on an organism or part of an organism.
Biramous: Divided into two branches.
Brackish water: Any mixture of sea water and fresh water with a salinity of less than 30 ppt (parts per thousand).
Branchiobdellids: Leechlike parasites of crayfish.
Branchiura: Fish lice.
Brine shrimp: Small crustacean (*Artemia* spp.) that can be easily reared as food for early stages of fishes and shellfishes.
Brood stock: Animals selected and maintained to reproduce and provide young for stocking grow-out facilities.
Bryozoans: Small marine and freshwater colonial animals that resemble mosses.
Byssus: Mass of horny fibers secreted by a gland in some mollusks; used for attachment.
Cages: Enclosures for animals with both the top and bottom covered; may be floating or standing on the substrate.
Cannibalistic: Eating its own species.
Carapace: Shell on the back of a turtle or crustacean.
Carnivorous: Feeding on flesh of animals.
Catadromous: Migrating from fresh water to salt water to spawn (e.g., eels).
Caudal: Pertaining to the tail.
Centrarchids: Members of the family that includes sunfish hybrids, crappies, pumpkinseeds, and large- and smallmouth bass.
Cephalothorax: In crustaceans, the body region formed by the fusion of head and thorax.
Cestode: Flat, ribbon-shaped worm, such as a tapeworm.
Chelate: Like a *chela* (pl. *chelae*); that is, pincer- or clawlike.
Cheliped: In crustaceans, a claw-bearing appendage.
Chitin: Horny, living substance forming the chief component of the exoskeleton of crustaceans.
Chitinoclastic bacteria: Microorganisms that destroy the chitinous exoskeleton of a number of different species of crustaceans; *cf.* shell disease.
Ciliated: Bearing *cilia*, hairlike outgrowths capable of whiplike beating movements.
Closed(-cycle) systems: System with no water-inflow requirements; water is treated to maintain good quality and then recirculated.
Coelenterates: Invertebrates that include the jellyfishes, hydras, sea anemones, and corals.
Cogeneration systems: Engines that burn hydrocarbons and produce both electricity and useful heat (i.e., hot water, steam, thermal energy).
Columnaris disease: Disease caused by a bacterium that enters the fish through the skin and gills.
Commensalism: Association between two species that is clearly to the advantage of one member but without seriously inconveniencing the other.
Competent larva: Larva of mollusk that is ready to metamorphose and attach to a suitable surface.
Competition: Struggle between organisms of the same (intraspecific) or different (interspecific) species for necessities of life.
Copepods: Small crustaceans, some species of which serve as food for many aquatic animals, while others are parasitic on the skin, scales, and gills of fish.
Coralline: Resembling a coral; containing or composed of coral.

Cotton-wool disease: Bacterial disease of fish manifested by off-white tufts around the mouth, fins, or body.

Crustaceans: Mainly aquatic arthropods characterized by an exoskeleton of chitin divided into head, thorax, and abdomen (e.g., shrimps, lobsters, copepods).

Crystalline style: In some mollusks, a translucent rod involved in carbohydrate digestion in the alimentary canal.

Cultch: Tiles, old oyster shells, or any material used by oyster farmers to collect the young of oysters (spat).

Cultivate: Raise crops through labor and care; to stir up oysters (on oyster beds), breaking up clusters and removing pests and predators.

Culture: Practice of cultivation, as of the soil or water; raising plants and animals with a view to their improvement.

Decapods: Order of crustaceans having five pairs of legs on the thorax; includes shrimps, crabs, and lobsters.

Demersal: Living near or on the bottom of the sea or lake.

Dentition: Number, type, and arrangement of teeth.

Depuration: Technique whereby animals that may contain undesirable substances (sand, silt), pollutants, parasites, or organisms of possible harm to humankind are placed and held in clean water to cleanse themselves; usually applied to shellfish.

Detritus: Fragments of dead and decomposing animals and plants; any waste.

Diadromous: Pertaining to fishes that migrate freely between salt and fresh waters.

Diatom: Microscopic unicellular or colonial algae with silica skeletons.

Digenea: Class of parasitic flukes or flatworms.

Digenetic trematodes: Parasitic flatworms with suckers and hooks; their life cycles usually require several hosts.

Dinoflagellates: Minute, mostly marine protozoans having two flagella; they form one of the chief constituents of plankton.

Diploid number: Number of paired chromosomes in a cell nucleus; twice the haploid number.

Disease: Deranged condition of an organism, whether inherited or caused by parasites, dietary deficiencies, or adverse physical and chemical factors in the environment.

Disinfection: System or procedure to control organisms that may cause harm to the cultured organism or to its end user. Methods of disinfection include heat, ultraviolet (UV) radiation, and chemicals.

Dorsal: Pertaining to the back surface of an animal.

Drills: Snails that destroy mollusks by rasping a hole in their shells and eating the fleshy portion.

Dropsy: Edema; excessive fluid in cells, tissues, or body cavities of fish caused by bacterial or viral infection.

Echinoids: Sea urchins and their allies.

Ecology: Study of the relations of organisms to their environment.

Ectoparasite: Organism that lives on another organism (the *host*) on which it is dependent for its own metabolism (e.g., leeches on the skin of fishes); *cf.* endoparasite.

Egtved disease: Viral disease of freshwater fish that infects the bloodstream.

Elver: Young eel.

Embryo: Early stage of an animal's development, usually before hatching from an egg.

Endemic: Peculiar to an area, as an endemic species of fish.

Endoparasite: Organism that lives inside another organism (the *host*) on which it is dependent for its metabolism; *cf.* ectoparasite.

Enteric redmouth: Highly infectious bacterial disease of trout and salmonids.

Epibiotic: Growing on the exterior of a living organism.

Epipelagic: Pertaining to that portion of the oceanic zone from the surface to about 600-ft depth into which enough light penetrates to allow photosynthesis.

Epiphyton: Plant that lives on the surface of another plant or animal.

Epizootic: Disease attacking a large number of animals nearly simultaneously (as an epidemic in humans).

Estuary: Generally areas of high productivity in bays or semienclosed areas where freshwater rivers and streams enter seawater.

Euryhaline: Adaptable to a wide range of salinity (applied to marine organisms).

Eutrophication: State of a water body with a good supply of nutrients and hence a rich organic production; has come to mean adverse or undesirable enrichment of a body of water.

Exoskeleton: External supportive covering of animals, such as scales of fishes or shells of crustaceans.

Fathom: Unit of length equal to 6 ft (~1.8 m).

Fecundity: Degree of fruitfulness, as indicated by the number of eggs produced by a female.

Filamentous: Threadlike; composed of filaments.

Filtration: Separation of a mixture or solution into component parts, which may be pure or mixtures of two or more constituents.

Fin rot: Split, ragged fins caused by bacterial infection.

Finfish: Cold-blooded lower aquatic vertebrates possessing fins and (usually) scales.

Fingerling: Young fish, larger than a fry but not an adult; up to 1 year old.

Fiord (fjord): Arm of the sea extending inland, usually long and narrow and bordered by steep cliffs.

Fish ranching: Releasing large numbers of young of anadromous species (e.g., salmon) into the sea to grow and return as adults to the area where they were released; sometimes called *extensive aquaculture.*

Fishery biologist: Person educated in fish biology and the management of fisheries.

Fishery economist: Person educated in economics who studies the ways best (on a monetary basis) to utilize exploited fish populations.

Flagellum: Long, whiplike structure (pl.: *flagella*) used for movement of an organism or to create currents that bring the necessities of life to it; generally longer and fewer in number than cilia; found in some protozoans and bacteria.

Food web (food chain): Transfer of food energy through a series of organisms via many stages of eating and being eaten (e.g., plants eaten by shrimps, shrimps eaten by fishes, and so on).

Forage species: Live feed for domestic species; animals that are placed or encouraged to grow in a pond to provide nourishment for a more desirable predatory species (e.g., sunfish in ponds with largemouth bass).

Foraminifera: Order of small, mostly marine protozoans with a shell usually perforated by pores.

Fry: Very young postlarval fish.

Furunculosis: Boils; skin abscesses.

Fusiform: Tapering at both ends; spindle-shaped.

Gas bubble disease: Bubbles of nitrogen in the blood and tissue of fish caused by supersaturation of the water by nitrogen gas.

Gastropods: Class of mollusks with a univalve shell or none; includes whelks, winkles, slugs, and snails.

Genetic engineering: Human manipulation of the genetic code to improve plants and animals.

Geothermal aquaculture: Use of warm water from wells naturally heated by the earth's interior to produce fish or shellfish commercially; permits raising warm-water species in colder climates.

Gonads: Reproductive organs of animals in which eggs and sperm are produced.

Gravid: Pregnant or ripe, ready to spawn.

Growing out: Maintaining crops until they reach harvest size.

Haploid number: Number of unpaired chromosomes in a cell nucleus; half the diploid number.

Hectocotylus: Arm of male cephalopod specialized for transfer of sperm to female.

Hemic: Pertaining to blood or blood vessels.

Hemorrhagic: Bleeding abnormally.

Hepatopancreas: Gland that secretes digestive enzymes.

Herbivorous: Feeding on plants.

Hermaphroditism: Having both male and female reproductive organs. In one type of hermaphroditism, the male and female organs are functional at the same time; in a second type, *sequential hermaphroditism,* a sex change occurs at some point in the life of the individual. Sequential hermaphroditism is called *protandry* when the initial sex is male and *protogyny* when the initial sex is female; the former is more common than the latter. Hermaphroditism occurs in 18 families of fishes and is widespread in invertebrates; example species of aquacultural interest include groupers and pandalid (northern) shrimp.

Hirudinea: Class comprising leeches, parasitic or predatory annelid worms.

Host: Organism on or in which a parasite lives, and from which it receives required metabolic products and, occasionally, other biological needs, such as protection from predators.

Hybrid: Offspring of two plants or animals from different species or varieties.

Hydrozoans: Class of coelenterates including simple and compound jellyfishes and polyps.

Hypersaline: Having a high salinity.

"Ich": Common name for a parasitic disease caused by the protozoan *Ichthyophthirius multifiliis* in fresh water and *Cryptocaryon irritans* in salt water.

Impounded: Confined within an enclosure.

Improve: Make useful or value-enhancing changes in (a species); *cf.* culture, genetic engineering.

Inbreeding: Mating of related individuals; this practice may improve races or strains of plants or animals.

Indicator organisms: Living organisms that indicate pollution in natural waters; placed in samples of water suspected to be polluted, survival of these organisms is tested.

Induced spawning: Egg-laying brought about by manipulation of the environment or treatment of the animal.

Instar: Stage between molts.

Interspecific competition: Competition among members of different species for the same resource.

Interspinous: Interspinal; between spines or spinal processes.

Intertidal zone: Area on the foreshore between lower low water and higher high water.

Intraspecific competition: Competition among members of the same species for the same resource.

Invertebrates: Lower animals, those without backbones.

Isopods: Small marine, freshwater, and terrestrial crustaceans with flattened bodies, fixed eyes, and no carapace; mostly scavengers.

Juveniles: Young stages of animals, usually between the postlarval stages up to the time they become sexually mature.

Keel: Narrow ridge.

Kelp: Large brown seaweeds (algae).
Lagoon: Shallow sound, pond, or lake, usually separated from the open ocean.
Larva: Immature stage of an animal that differs greatly in appearance and behavior from a juvenile or adult.
Leeches: *See* Hirudinea.
Leptocephalus: Translucent larval stage of some eels, before the elver stage.
Lernaea: Genus comprising anchor worms, freshwater parasites that usually invade the skin or blood vessels of fish.
Lesions: Sores or open wounds; may be caused by microorganisms or parasites.
Littoral zone: Shallow, inshore areas, usually of depths down to 200 fathoms.
Lymphocystis disease: Viral infection of fish, producing growths on the fins.
Lymphosarcoma: Malignant infection of the lymph tissues.
Mariculture: *See* sea farming.
Megalops: Larval stage of certain crustaceans, such as crabs.
Meiofauna: Small animals that live near, on, or in the bottom deposits of a body of water.
Mesenteries: Membranes that hold viscera in position.
Mesozoa: Small wormlike animals that are parasitic on marine organisms.
Metamorphosis: Abrupt physical change in form (e.g., from tadpole into frog).
Microencapsulation: Artificial feed encapsulated within a biodegradable wall designed to protect the food from bacteria while allowing leaching of nutrients.
Microsporeans: Intracellular parasites with spores that cause cysts or tumors.
Mollusks: Animals with soft body coverings and limy shells of 1–18 parts or sections; in some species the shell is lacking or reduced in size.
Molt: One of a number of periodic sheddings of the outer covering, such as the exoskeleton in the arthropods (shrimps, crabs, etc.).
Molt death syndrome: Dietary deficiency preventing American lobsters from molting; usually fatal.
Monogenea: Class of parasitic flatworms or flukes that are principally ectoparasites of fish.
Monogenetic trematodes: Ectoparasitic flatworms with hooks and suckers; their life history involves no intermediate hosts.
Monosex culture: Raising a single sex in an impoundment to prevent breeding; sexes may have to be sorted or bred from hybrids that produce young of a single sex.
Mutations: Changes in genes that cause new characters to appear in a species.
Mycobacteriosis: Fish tuberculosis.
Mysis: Last larval stages in crustaceans before they transform into juveniles.
Myxosporeans: Class of protozoans that produce spores parasitic to some fish and invertebrates.
Nauplius: First larval stage occurring in many crustacean species (e.g., shrimps): characterized by an unsegmented body and three pairs of appendages.
Necrosis: Death of a cell or group of cells due to injury or disease.
Nematodes: Roundworms that are both free-living and parasitic in plants and animals; they occur in many species of aquatic animals.
Nemertea: Class of long, slender marine worms that mostly live on shores along the low-tide line.
Neoplasm: Uncontrolled cell growth; tumor.
Net pens: Large enclosures usually placed in protected bays, fjords, or inlets and supported near or at the sea surface by floats. Netting covers the bottom of the net pen, but the top is usually uncovered, except to prevent predation on the enclosed animals or their escape by jumping; *cf.* cages.

Off-flavor: Environment-related flavors, like "muddy" or "musty," permeating pond-raised fish.

Oligochaetes: Order of annelids (segmented worms) that includes earthworms and related species that live in marine and fresh water.

Omnivorous: Feeding on both plant and animal matter.

Open(-cycle) system: System that depends on good-quality incoming water, which is discharged when the water properties are unacceptable.

Opercular: In fishes, posterior bone of the gill cover.

Operculum: Lidlike structure that closes the mouth of a shell.

Orbicular: Round, shaped like a sphere.

Organoleptic: Perceived by, or pertaining to, a sensory organ.

Ozone: Form of oxygen that can be used as a water purifier.

Parasite: Organism that lives part or all of its life in or on another organism (the *host*) on which it is dependent for its metabolism.

Paratonic: Retarding or checking movement or growth.

Pathogenic: Causing disease.

Pediveliger: Larval stage of oysters that still has the swimming ciliated organ (*velum*) and sensitive foot needed for settlement and attachment.

Peduncle: Stem or stalk; a narrow part by which some larger part (or the whole body) is attached.

Pelagic: Living in the middle or surface levels of the sea; usually used to describe high-seas fish that migrate widely (e.g., tunas, salmon).

Pentastomida: Class of bloodsucking parasitic arthropods.

Pesticide: Chemical substance used to kill pests.

pH: Scale of 0–14 that indicates the acidity ($pH < 7$) and alkalinity ($pH > 7$) of solutions; a value of 7 means the solution is neutral.

Pharyngeal: Pertaining to the throat.

Photoperiodism: Physiological responses to the length of night or day, or both.

Phytoplankton: Plant plankton; primarily microscopic, largely diatomic plants that drift with the currents and, together with *zooplankton,* make up the *plankton,* which are the basic synthesizers of organic matter.

Plankton: Passively drifting or weakly swimming organisms in fresh water and the sea, including many microscopic plants and animals, but also some large jellyfish; plant plankton are called *phytoplankton;* animal plankton, *zooplankton.*

Plastron: Bony plate on the underside of turtles.

Pleopod: *See* swimmeret.

Pollution: Specific impairment of water quality by sewage, pesticides, and industrial wastes; may create a hazard to public health or kill or stress aquatic organisms.

Polychaetes: Marine worms with paired, flattened, bristle-tipped organs of locomotion.

Polyculture: Raising two or more noncompetitive species in the same place.

Polyphagous: Eating various kinds of food.

Polyploidy: Condition where the number of chromosomes in the body cells is three or more times the haploid number. Can be induced in some aquacultural species by reduced temperatures or increasing pressures, causing sterility and thereby increasing the animal's growth rate. Used also for the introduction of exotic species, such as the grass carp, so that their numbers can be kept in check and thus not compete with local species.

Postlarvae: Stages, past the larval stage, that resemble the juvenile but still lack certain characters.

Predation: Act of an animal eating another (prey) of a different, and usually smaller, species.

Prey: Animal that is hunted and eaten by predatory animals.

Propagation: Multiplication of plants and animals by any method from parent stock.

Protandry: Condition in sequential hermaphrodite species where the male reproductive cells mature and are shed before the female reproductive cells mature.

Protogyny: Condition in sequential hermaphrodite organisms in which the female reproductive structures mature before the male structures.

Protozoans: Single-celled animals.

Protozoea: Larval stages between the *nauplius* and *mysis* in crustaceans; usually have seven pairs of appendages.

Raceway: Culture chamber, usually long and narrow, in which the water enters one end and exits the other.

Radula: Horny band or ribbon in mollusks (other than bivalves); minute teeth on its dorsal surface tear up food and draw it into the mouth.

Raft culture: Growing oysters or mussels on shells or other materials suspended from rafts or floats; sometimes used to describe any method of hanging culture.

Raise (rear): Cause or promote the growth of crops (plants or animals).

Raptorial predator: Predator adapted for seizing their prey (e.g., some copepods).

Recirculation: Reuse of water in an aquaculture facility (*closed* system) rather than releasing into nature and continually replaced by new water (*open* system).

Relay: Collecting oysters, clams, or mussels in one location and planting them in another to obtain better growth or better quality meats.

Remote setting: Shipping of cultured larval oysters, still in the *pediveliger* stage, for final settlement on suitable substrate (*cultch*).

Resistant: Able to withstand adverse environmental conditions or ward off diseases.

Rickettsia: Family of bacterial cells often found in arthropods; includes human and animal pathogens and parasites.

Riparian rights: Rights concerning the bank, bed, or waters belonging to a person who owns land bordering on a watercourse or other body of water.

Roe: Eggs of aquatic fishes when still enclosed in the ovarian membrane; in the case of sea urchins, both male and female gonads are marketed as "roe."

Rostrum: Beak or beaklike structure.

Rotifer: Minute, microscopic, multicellular animal living mostly in fresh water; usually conical with a crown of hairlike outgrowths at the widest end that seems to be rotating.

Salinity: Measure of the quantity of dissolved salts in seawater; the total amount of dissolved solids in seawater in parts per thousand (ppt, or ‰) by weight when all the carbonate has been converted to oxide, the bromide and iodide to chloride, and all organic matter is completely oxidized. Salinity in the open sea varies from about 33 to 37 ppt. Estuaries receiving freshwater runoff have reduced salinities; lagoons with limited exchange with the open ocean may have highly elevated salinities.

Salmonid: Any of a family of elongate soft-finned fishes that have the last verterbrae upturned (e.g., salmon, trout).

Scute: External scale or horny plate on some fishes and many reptiles.

Sea farming: Promote or improve growth and hence production of marine and brackish-water plants and animals by labor and attention (husbandry), at least at some stage of the life cycle, on areas leased or owned; usually intended as a profit-making venture.

Seed: Young animals—generally oysters, clams, or mussels—used for stocking.

Seawater: Water usually with salinity of 30–37 ppt, as found in the open oceans; the salinity of estuarine waters usually varies around this value.

Seawater system: System intended for the culture or maintenance of marine organisms; there are several types of both *open*- and *closed*-cycle systems.

Segmentation: Division into parts or segments, as in annelid worms.

Sessile: Permanently fixed or attached; not free-moving.

Shell disease syndrome: Pitting and erosion of the exoskeleton of marine and freshwater crustaceans, believed to be caused by chitin-destroying bacteria and fungi of several genera; will result in death in stressed animals.

Shellfishes: Aquatic invertebrates possessing a shell or exoskeleton; usually mollusks or crustaceans.

Shuck: Remove shells (from oysters, clams, etc.) for market, or in preparation for eating.

Silos: Deep circular tanks or "vertical raceways," requiring a relatively large flow of high-quality water in order to provide a gravitational head sufficient to flush wastes from the tanks. Disadvantages include difficulty of grading fish to control cannibalism and of harvesting just a portion of the fish from the silo.

Siphon: Funnel-shaped structure in aquatic mollusks via which water is drawn in and out.

Spat: Young oysters just past the veliger stage; those that have settled down and become attached to some hard object.

Spat-fall: Settling down or attachment of young oysters that have completed their larval stages.

Spawn: Deposit or produce eggs, sperm, or young.

Specificity: Range of hosts in or on which individual species of parasite will live.

Spermatophore: Small packet of sperm cells produced by some species of animals.

Spinning disease: Loss of nervous control that causes a squid to spin; may be due to lack of strontium in the rearing water.

Spinule: Small spine.

Stomatopoda: Order of marine crustaceans with raptorial arms; includes the mantis shrimp.

Sublittoral: Shallow; pertaining to a water zone down to about 200 m in depth.

Substrate: Material (sand, mud, etc.) that covers the bottom of an aquatic environment.

Subterminal: Near the end.

Subtropical zone: Area just outside the tropics (23.5° N and S of the equator) with a climate similar to that of the tropics.

Suctoria: Parasitic ciliate that attacks the gills of fish.

Superinfection: Reinfection by parasites in a host already harboring these parasites.

Swimmeret: Small, paired crustacean appendage involved partly in swimming.

Symbiont: One of the organisms in symbiosis.

Symbiosis: Relationship of two or more different organisms in close association that may be (but is not necessarily) of mutual benefit.

Telson: Unpaired terminal segment on the abdomen of crustaceans.

Temperate zone: Region from 23.5° N and S of the equator to the respective polar regions; areas with definite seasons.

Test: Hard outer coating or shell.

Tidal marshes: *See* wetlands.

Transgenic species: Animal or plant into which genes from another species have been deliberately introduced by genetic engineering.

Trematodes: Group of parasitic flatworms, some of which occur as parasites of aquatic animals.

Trichodina: Common ectoparasitic ciliate that may be pathogenic to aquatic organisms.

Triploid: Organism or cell treated by laboratory techniques to produce three sets of chromosomes per nucleus and, thereby, sterile individuals. Such sterilization permits species like grass carp to be released in nature without the danger of overreproduction and possibe harm to the ecology of an area.

Trochophore: Top-shaped, free-swimming larva of some mollusks and annelids.

Tubercles: Small rounded nodules or swellings.

Tunicates: Highly modified marine chordates, usually globular or cylindrical in shape, and having a secretion of outer covering, or tunic, about the body.

Turbellaria: Class of free-living flatworms.

Turbidity: Cloudy condition of water, usually caused by impurities; may result from wave action stirring up bottom sediments.

Ultraviolet (UV) light: Electromagnetic radiation that lies just beyond the violet end of the visible spectrum. The radiation produced by this light is used in aquaculture to disinfect water and prevent diseases caused by pathogenic microorganisms.

Upwelling: Process by which nutrient-rich bottom waters of the sea are brought near the surface.

Uropods: Fan-shaped paired legs on one of the body segments of crustaceans; used for swimming.

Veliger: Molluscan larval stage that entails a band of small hairs (*cilia*) that aids in weak locomotion.

Velum: Ciliated swimming organ developed in later larval stages of many marine gastropods.

Ventral: Pertaining to the belly or underside.

Vertebrates: Higher animals; those with backbones (vertebrae).

Vibrio: Genus of bacterial pathogens of marine fishes.

Wetlands: Areas near the sea that are alternatively flooded and drained by tidal action; also called tidal marshes.

Whirling disease: Highly infectious parasitic disease that causes salmonids to swim erratically.

Whorl: Spiral turn of a shell.

Zoea: Larval stage of some arthropods, such as shrimp and crabs.

Zooplankton: Animal plankton, such as crustaceans and jellyfishes, that drift with the currents, consume phytoplankton and other zooplankton, and serve as food for large animals.

Zoospore: Motile, usually naked and flagellated, asexual spore, especially of an alga or lower fungus.

Zooxanthellae: Microscopic yellow or brown algae that live symbiotically in various animals.

Appendix B

Major Groups of Aquacultural Species

AQUATIC PLANTS

Microalgae
Spirulina
Tetraselmis
Isochrysis

Macroalgae
Irish moss
Giant kelp
Nori

Aquatic plants
Water hyacinth

LIVE ROCK

(Dead coral and noncoral limestone permeated with bacteria, encrusted with algae, and home to a variety of invertebrates)

INVERTEBRATES

Annelids
Bloodworm
Lugworm
Sandworm

Arthropods/Crustaceans
Brine shrimp
Crab
Copepods
Crayfish
Daphnia
Lobster
Shrimp/prawn

Echinoderms
Sea urchins

Mollusks
Bivalves
- Clam
- Mussel
- Pen shell
- Oyster
- Scallop

Gastropods
- Abalone
- Conch

Cephalopods
- Octopus
- Squid

Rotifers

Sponges

VERTEBRATES

Amphibians

Frog

Fishes

American shad
Arctic char
Atlantic salmon
Bowfin
Buffalo fish
Bull minnow (bait)
Channel catfish
Cod (Atlantic)
Coho salmon
Common carp
Crappie
Croaker
Dolphin fish
Eel
Flatfish
Flounder
Golden shiner (bait)
Goldfish (ornam.)
Grass carp
Grouper
Halibut (Atlantic)
King salmon
Koi (ornam.)
Largemouth bass
Milkfish
Mullet (striped)
Muskellunge
Paddlefish
Pike
Pink salmon
Plaice
Pompano (Florida)
Pumpkinseed
Red drum (redfish)
Seatrout (weakfish)
Smallmouth bass
Snapper
Snook
Sole
Steelhead (rainbow) trout
Striped bass
Sturgeon
Sunfish hybrids (bluegill, bream)
Threadfin shad
Tilapia
Tuna
Turbot
Walking catfish
Walleye
Yellow perch
Yellowtail

Reptiles

Alligator
Crocodile
Turtle (incl. terrapin)

Appendix C

Aquacultural Species Common Name/Scientific Name

Common Name	Scientific Name
Abalone	*Haliotis rufescens*
American alligator	*Alligator mississippiensis*
American bullfrog	*Rana catesbeiana*
American crocodile	*Crocodylus acutus*
American eel	*Anguilla rostrata*
American lobster	*Homarus americanus*
American oyster	*Crassostrea virginica*
Apple snail	*Ampullaria*
Apple snail	*Pila* spp.
Apple snail	*Pomacea* spp.
Arctic char	*Salvelinus alpinus*
Atlantic cod	*Gadus morhua*
Atlantic croaker	*Micropogonias undulatus*
Atlantic halibut	*Hippoglossus hippoglossus*
Atlantic salmon	*Salmo salar*
Bay scallop	*Argopecten irradians*
Bigmouth buffalo	*Ictiobus cyprinellus*
Black buffalo	*Ictiobus niger*
Black crappie	*Pomoxis nigromaculatus*
Black grouper	*Mycteroperca bonaci*
Bloodworm	*Glycera dibranchiata*
Blue crab	*Callinectes sapidus*
Bluefin tuna	*Thunnus thynnus*
Bluegill	*Lepomis macrochirus*
Bowfin	*Amia calva*
Brine shrimp	*Artemia salina*
Brook trout	*Salvelinus fontinalis*
Bull minnow	*Fundulus grandis*
Carp	*Cyprinus carpio*
Channel catfish	*Ictalurus punctatus*
Chinook salmon	*Oncorhynchus tshawytscha*
Chub sucker	*Erimyzon* spp.
Coho salmon	*Oncorhynchus kisutch*
Copepod	*Acartia* spp.
Diamond-back terrapin	*Malaclemys* spp.
Dolphin fish	*Coryphaena hippurus*
Dungeness crab	*Cancer magister*
European eel	*Anguilla anguilla*
European lobster	*Homarus gammarus*
Flathead minnow	*Pimephales promelas*
Geoduck clam	*Panope abrupta (generosa)*
Giant clam	*Hippopus hippopus*
Giant clam	*Tridacna derasa*
Giant clam	*Tridacna gigas*
Giant clam	*Tridacna squamosa*
Giant kelp	*Macrocystis pyrifera*
Giant river prawn	*Macrobrachium rosenbergii*
Giant tiger prawn	*Penaeus monodon*
Golden shiner	*Notemigonus crysoleucas*
Goldfish	*Carassius auratus*
Grass carp	*Ctenopharyngodon idella*
Grass shrimp	*Palaemonetes* spp.
Grass sponge	*Spongia graminea*
Great land crab	*Cardisoma guanhumi*
Great white lucine	*Codakia orbicularis*
Green mussel	*Perna viridis*
Green sea turtle	*Chelonia mydas*
Green sunfish	*Lepomis cyanellus*
Hard clam	*Mercenaria mercenaria*
Hermit crab	*Coenobita clypeatus*
Irish moss	*Chondrus crispus*
Killifish	*Fundulus* spp.
Koi	*Cyprinus carpio*
Kuruma prawn	*Penaeus japonicus*

Common name	Scientific name
Largemouth bass	*Micropterus salmoides*
Lugworm	*Arenicola cristata*
Mangrove oyster	*Crassostrea rhizophorae*
Manila clam	*Tapes japonica*
Marron	*Cherax tenuimanus*
Microalgae	*Spirulina* spp.
Milkfish	*Chanos chanos*
Mullet	*Mugil cephalus*
Muskellunge	*Esox masquinongy*
Nori	*Porphyra* spp.
Octopus	*Octopus joubini*
Pacific top shell	*Trochus niloticus*
Pacific white shrimp	*Penaeus vannamei*
Paddlefish	*Polyodon spathula*
Paddlefish	*Psephurus gladius*
Pearl oyster	*Pinctada martensii*
Pen shell	*Atrina rigida*
Pen shell	*Pinna carnea*
Pike	*Esox lucius*
Pink salmon	*Oncorhynchus gorbuscha*
Pompano	*Trachinotus carolinus*
Pompano dolphin	*Coryphaena equisetis*
Pumpkinseed	*Lepomis gibbosus*
Queen conch	*Strombus gigas*
Rainbow trout	*Oncorhynchus mykiss*
Red drum	*Sciaenops ocellatus*
Red-eared terrapin	*Trachemys scripta*
Red grouper	*Epinephelus morio*
Red sea urchin	*Strongylocentrotus franciscanus*
Red snapper	*Lutjanus campechanus*
Red swamp crawfish	*Procambarus clarkii*
Rock scallop	*Crassadoma gigantea*
Rotifer	*Brachionus plicatilis*
Sandworm	*Nereis* spp.
Sea hare	*Aplysia californica*
Sea mussel	*Mytilus edulis*
Sea scallop	*Argopecten circularis*
Sea scallop	*Argopecten purpuratus*
Sea scallop	*Patinopecten yessoensis*
Sea urchin	*Anthocidaris crassispina*
Sea urchin	*Hemicentrotus pulcherrimus*
Sea urchin	*Pseudocentrotus depressus*
Shiner	*Notropis* spp.
Smallmouth bass	*Micropterus dolomieu*
Smallmouth buffalo	*Ictiobus bubalus*
Snook	*Centropomus undecimalis*
Soft-shelled clam	*Mya arenaria*
Spiny lobster	*Panulirus argus*
Spotted seatrout	*Cynoscion nebulosus*
Squid	*Illex* spp.
Squid	*Loligo* spp.
Steelhead	*Oncorhynchus mykiss*
Stone crab	*Menippe mercenaria*
Stone roller	*Campostoma* spp.
Striped bass	*Morone saxatilis*
Surf clam	*Spisula solidissima*
Threadfin shad	*Dorosoma petenense*
Tiger lucine	*Asaphis deflorata*
Tilapia	*Sarotherodon* spp.
Tilapia	*Tilapia mossambica*
Top minnow	*Poecilia* spp.
Tubifex worm	*Branchiura sowerbyi*
Tubifex worm	*Limnodrilus* spp.
Tubifex worm	*Lumbriculus variegatus*
Tubifex worm	*Tubifex tubifex*
Unicellular microalgae	*Isochrysis* spp.
Unicellular microalgae	*Tetraselmis* spp.
Walking catfish	*Clarias batrachus*
Walleye	*Stizostedion vitreum*
Water flea	*Daphnia* spp.
Water flea	*Moina* spp.
Water hyacinth	*Eichhornia crassipes*
West Indian king crab	*Mithrax spinosissimus*
West Indian top shell	*Cittarium pica*
White bass	*Morone chrysops*
White crappie	*Pomoxis annularis*
White river crawfish	*Procambarus blandingii*
White sturgeon	*Acipenser transmontanus*
Wool sponge	*Hippiospongia lachne*
Yellow perch	*Perca flavescens*
Yellow sponge	*Spongia barbara*
Yellowtail	*Seriola quinqueradiata*

Appendix D

Aquacultural Species Scientific Name/Common Name

Scientific Name	Common Name
Acartia spp.	Copepod
Acipenser transmontanus	White sturgeon
Alligator mississippiensis	American alligator
Amia calva	Bowfin
Ampullaria spp.	Apple snail
Anguilla anguilla	European eel
A. rostrata	American eel
Anthocidaris crassispina	Sea urchin
Aplysia californica	Sea hare
Arenicola cristata	Lugworm
Argopecten circularis	Sea scallop
A. irradians	Bay scallop
A. purpuratus	Sea scallop
Artemia salina	Brine shrimp
Asaphis deflorata	Tiger lucine
Atrina rigida	Pen shell
Brachionus plicatilis	Rotifer
Branchiura sowerbyi	Tubifex worm
Callinectes sapidus	Blue crab
Campostoma spp.	Stone roller
Cancer magister	Dungeness crab
Carassius auratus	Goldfish
Cardisoma guanhumi	Great land crab
Centropomus undecimalis	Snook
Chanos chanos	Milkfish
Chelonia mydas	Green sea turtle
Cherax tenuimanus	Marron
Chondrus crispus	Irish moss
Chrysemys scripta, see *Trachemys scripta*	
Cittarium pica	West Indian top shell
Clarias batrachus	Walking catfish
Codakia orbicularis	Great white lucine
Coenobita clypeatus	Hermit crab
Coryphaena equisetis	Pompano dolphin
C. hippurus	Dolphin fish
Crassadoma gigantea	Rock scallop
Crassostrea rhizophorae	Mangrove oyster
C. virginica	American oyster
Crocodylus acutus	American crocodile
Ctenopharyngodon idella	Grass carp
Cynoscion nebulosus	Spotted seatrout
Cyprinus carpio	Carp
Daphnia spp.	Water flea
Dorosoma petenense	Threadfin shad
Eichhornia crassipes	Water hyacinth
Epinephelus morio	Red grouper
Erimyzon spp.	Chub sucker
Esox lucius	Pike
E. masquinongy	Muskellunge
Fundulus spp.	Killifish
F. grandis	Bull minnow
Gadus morhua	Atlantic cod
Glycera dibranchiata	Bloodworm
Haliotis rufescens	Abalone
Hemicentrotus pulcherrimus	Sea urchin
Hinnites multirugosus, see *Crassadoma gigantea*	
Hippiospongia lachne	Wool sponge
Hippoglossus hippoglossus	Atlantic halibut
Hippopus hippopus	Giant clam
Homarus americanus	American lobster
H. gammarus	European lobster
H. vulgaris, see *Homarus gammarus*	
Ictalurus punctatus	Channel catfish
Ictiobus bubalus	Smallmouth buffalo
I. cyprinellus	Bigmouth buffalo
I. niger	Black buffalo
Isochrysis spp.	Unicellular microalgae
Lepomis gibbosus	Pumpkinseed

L. macrochirus	Bluegill
L. cyanellus	Green sunfish
Limnodrilus spp.	Tubifex worm
Loligo spp.	Squid
Lumbriculus variegatus	Tubifex worm
Lutjanus campechanus	Red snapper
Macrobrachium rosenbergii	Giant river prawn
Macrocystis pyrifera	Giant kelp
Malaclemys spp.	Diamond-back terrapin
Menippe mercenaria	Stone crab
Mercenaria mercenaria	Hard clam
Micropogonias undulatus	Atlantic croaker
Micropterus dolomieu	Smallmouth bass
M. salmoides	Largemouth bass
Mithrax spinosissimus	West Indian king crab
Moina spp.	Water flea
Morone chrysops	White bass
M. saxatilis	Striped bass
Mugil cephalus	Mullet
Mya arenaria	Soft-shelled clam
Mycteroperca bonaci	Black grouper
Mytilus edulis	Sea mussel
Nereis	Sandworm
Notemigonus crysoleucas	Golden shiner
Notropis spp.	Shiner
Octopus joubini	Octopus
Oncorhynchus gorbuscha	Pink salmon
O. kisutch	Coho salmon
O. mykiss	Steelhead trout
O. tshawytscha	Chinook salmon
Palaemonetes pugio	Grass shrimp
P. vulgaris	Grass shrimp
P. intermedius	Grass shrimp
Panope abrupta (generosa)	Geoduck
Panulirus argus	Spiny lobster
Patinopecten yessoensis	Sea scallop
Penaeus japonicus	Kuruma prawn
P. monodon	Giant tiger prawn
P. vannamei	Pacific white shrimp
Perca flavescens	Yellow perch
Perna viridis	Green mussel
Pila spp.	Apple snail
Pimephales promelas	Flathead minnow
Pinctada martensii	Pearl oyster
Pinna carnea	Pen shell
Poecilia spp.	Top minnow
Polyodon spathula	Paddlefish
Pomacea spp.	Apple snail
Pomoxis annularis	White crappie
P. nigromaculatus	Black crappie
Porphyra spp.	Nori
Procambarus blandingii	White river crawfish
P. clarkii	Red swamp crawfish
Pseudocentrotus depressus	Sea urchin
Psephurus gladius	Paddlefish
Rana catesbeiana	American bullfrog
Salmo gairdneri, see Oncorhynchus mykiss	
S. salar	Atlantic salmon
Salvelinus alpinus	Arctic char
S. fontinalis	Brook trout
Sarotherodon spp.	Tilapia
Sciaenops ocellatus	Red drum
Seriola quinqueradiata	Yellowtail
Spirulina spp.	Microalgae
Spisula solidissima	Surf clam
Spongia barbara	Yellow sponge
S. graminea	Glass sponge
Stizostedion vitreum	Walleye
Strombus gigas	Queen conch
Strongylocentrotus franciscanus	Red sea urchin
Tapes japonica	Manila clam
Tetraselmis spp.	Unicellular microalgae
Thunnus thynnus	Bluefin tuna
Tilapia mossambica	Tilapia
Trachemys scripta	Red-eared terrapin
Trachinotus carolinus	Pompano
Tridacna derasa	Giant clam
T. gigas	Giant clam
T. squamosa	Giant clam
Trochus niloticus	Pacific top shell
Tubifex tubifex	Tubifex worm

Appendix E

Geographic Listing of Major and Potential North American Aquacultural Species

This list provides guidance to individuals seeking suitable candidate species for a particular North American region. It includes four kinds of species:

- those farmed commercially;
- those used in government fish-rearing programs for augmenting wild populations;
- those untried but that appear to be good candidates for aquaculture; and
- those that were tried and the ventures abandoned.

To obtain a complete list of aquaculture species, contact state fisheries and/or aquaculture agencies. Mailing addresses are provided for all extant state aquaculture coordinators.

A gross breakdown of North American aquacultural production is presented in Table E.1.

U.S. STATES AND POSSESSIONS

ALABAMA: *Freshwater*—bass (largemouth, striped), carp (common, grass), catfish, paddlefish, spotted seatrout; *Marine*—Florida pompano, red drum; *Coordinator*—Alabama Farmers Market Authority, P.O. Box 3336, Montgomery, AL 36193.

ALASKA: *Marine*—clams (littleneck), mussels, oysters, sablefish, salmon (state and federal hatcheries), scallops, seaweeds.

ARIZONA: *Freshwater*—bait fish (goldfish, minnows, shiners), bass (incl. largemouth), bluegills, catfish, crayfish, frogs, prawns, red drum, tilapia, trout (brown, rainbow).

ARKANSAS: *Freshwater*—bait fish (goldfish, minnows, shiners), bass (incl. striped), bluegills, buffalo fish, carp (common, grass), catfish, crappies, crayfish, tilapia, trout (rainbow); *Coordinator*—Arkansas Grain Warehouse & Catfish Proc. Sec., P.O. Box 1069, c/o Arkansas State Plant Board, Little Rock, AR 72203.

CALIFORNIA: *Freshwater*—algae, alligator, bait fish (goldfish, minnows, shiners), bass (largemouth, striped), bluegills, catfish, carp (common, grass, koi), crappies, crayfish, frogs, giant river prawns, guppies, mosquito fish, pearl mussels, sturgeon, sunfish, tilapia, tubificid worms, trout (brook, brown, rainbow); *Marine*—abalone, brine shrimp (*Artemia*, hypersaline), California halibut, mussels, oysters, salmon, shrimp (*Penaeus*), sturgeon, white sea bass.

COLORADO: *Freshwater*—carp, crayfish, tropical fish, trout.

CONNECTICUT: *Marine*—clams (surf), salmon (Atlantic), scallops.

DELAWARE: *Freshwater*—bass (striped), catfish, eel, trout; *Marine*—clams, oysters.

FLORIDA: *Freshwater*—alligators, aquatic plants, bass (incl. hybrid striped), bream, carp (grass, koi), catfish, crappie (black), crayfish, giant river prawns, ornamentals, red drum, sturgeon, tilapia, tropical fish; *Marine*—clams, groupers, killifish, mullet, oysters, red drum, sea hare (sea slugs), seahorses, snapper, spotted seatrout, yellowtail; *Coordinator*—Florida Dept. of Agriculture & Consumer Services, Room 425, Mayo Bldg., Tallahassee, FL 32399.

GEORGIA: *Freshwater*—bait fish (goldfish, minnows), bass (striped), bream, catfish, shad (American), suckers, tilapia, trout (rainbow), yeast fungi.

GUAM: *Marine*—catfish (Asian), giant clams, milkfish, oysters, rabbitfish, seaweed, tilapia.

HAWAII: *Freshwater*—carp (grass), catfish, frogs, giant river prawn, tilapia; *Marine*—algae (nori), abalone, lobsters, mussels, mahimahi (dolphin fish), milkfish, mullet, pearl oysters, shrimp (*Penaeus*); *Coordinator*—Aquaculture Development Program, State of Hawaii, Dept. of Land & Natural Resources, 335 Merchant St., Room 348, Honolulu, HI 96813.

IDAHO: *Freshwater*—bass (largemouth), bluegills, carp (grass), catfish, crappie, pike (walleye), salmon, snails, tilapia, tropical fish, trout (incl. rainbow); *Coordinator*—Idaho Dept. of Agriculture, Div. of Animal Industries, Box 7249, Boise, ID 83702.

ILLINOIS: *Freshwater*—bait fish, bass (incl. striped), catfish; *Coordinator*—Illinois Dept. of Agriculture, Aquaculture Specialist, State Fairgrounds, P.O. Box 19281, Springfield, IL 62794.

INDIANA: *Freshwater*—bass (largemouth), carp, catfish; *Coordinator*—Dept. of Animal Sciences, Purdue University, Poultry Bldg., W. Lafayette, IN 47907.

IOWA: *Freshwater*—bass (largemouth, striped), bluegill, carp (grass), catfish, crappie, muskellunge, paddlefish, pike (walleye), trout.

KANSAS: *Freshwater*—bass (largemouth, striped), bluegills, carp (common, grass), catfish, crappie, minnows, pike (walleye); *Coordinator*—Agricultural Marketing Programs Coordinator, 901 S. Kansas Ave., Topeka, KS 66612.

Table E.1. North American Aquacultural Production, 1986

	Production (mt)	% of Total[a]
Finfish[b]	262,149	49.5
Mollusks[b]	186,664	35.2
Crustaceans	80,863	15.3
Seaweeds	210	0.04
Total	529,886[c]	100

[a]Percentages are rounded.
[b]*FAO Aquaculture Minutes* (no. 8, 6 August 1990) lists Pacific salmon, catfish, and mollusks as key U.S. aquacultural products.
[c]Total world aquacultural production in 1988 was 13,708,963 mt, of which U.S. production comprised <5%.
Source: FAO data (metric tons).

KENTUCKY: *Freshwater*—bass (largemouth, striped), catfish, muskellunge; *Coordinator*—Marketing Supervisor, Kentucky Dept. of Agriculture, #63 Wilkinson Blvd., Frankfort, KY 40601.

LOUISIANA: *Freshwater*—alligator, catfish, crayfish, frogs, giant prawns, pearl mussels, tropical fish; *Marine*—red drum, shrimp (*Penaeus*); *Coordinator*—Louisiana Dept. of Agriculture & Forestry, Office of Marketing, P.O. Box 3334, Baton Rouge, LA 70821.

MAINE: *Freshwater*—trout; *Marine*—bluefin tuna, eels, flounders, mussels, rainbow smelt, salmon (Atlantic, Pacific).

MARYLAND: *Freshwater*—aquatic plants, carp (koi), catfish, goldfish, pike (walleye), trout; *Marine*—bass (hybrid striped), clams, crabs (soft-shell), oysters, tilapia.

MASSACHUSETTS: *Freshwater*—bass (largemouth), tilapia; *Marine*—clams, mussels, scallops.

MICHIGAN: *Freshwater*—bait fish (goldfish, shiners), bass (largemouth), bluegills, catfish, muskellunge, perch (incl. yellow), pike (northern, walleye), shad (gizzard), snails, suckers, sunfish, sturgeon, tilapia, trout (brook, brown, rainbow); *Coordinator*—Michigan Dept. of Agriculture, Director of the Center for Agric. Innovation & Development, P.O. Box 30017, Lansing, MI 48909.

MINNESOTA: *Freshwater*—bait fish (fathead minnow, shiners), bass (largemouth), buffalo fish (bighead), bullhead, carp, catfish (channel), crappie, crayfish, leech, muskellunge, perch (yellow), pike (northern, walleye), salmon (Atlantic, Pacific), suckers, sunfish, sturgeon, tilapia, whitefish; *Coordinator*—Minnesota Dept. of Agriculture, Aquaculture Coordinator, 90 W. Plato Blvd., St. Paul, MN 55107.

MISSISSIPPI: *Freshwater*—bass (striped), catfish, crayfish; *Marine*—oysters, red drum; *Coordinator*—Dept. of Agriculture & Commerce, Director of Marketing, P.O. Box 1609, Jackson, MS 39215.

MISSOURI: *Freshwater*—bass (striped), carp, catfish, muskellunge, paddlefish, trout (rainbow); *Coordinator*—Missouri Dept. of Agriculture, Market Development Div., P.O. Box 630, Jefferson City, MO 65102.

MONTANA: *Freshwater*—bass, minnows, paddlefish, salmon, sturgeon, trout (rainbow); *Coordinator*—Montana Dept. of Agriculture, Marketing Program Manager, Agriculture/Livestock Bldg., Capitol Station, Helena, MT 57620.

NEBRASKA: *Freshwater*—bait fish, bass (largemouth).

NEVADA: *Freshwater*—catfish, frogs, shad (threadfin), trout.

NEW HAMPSHIRE: *Freshwater*—trout; *Coordinator*—New Hampshire Dept. of Agriculture, Caller Box 2042, Concord, NH 03302.

NEW JERSEY: *Marine*—clams, oysters, shrimp; *Coordinator*—New Jersey Dept. of Agriculture, Fish & Seafood Promotion Technician, CN 330, Trenton, NJ 08625.

NEW MEXICO: *Freshwater*—catfish, Colorado squawfish, trout.

NEW YORK: *Freshwater*—carp, minnows, perch, pike (walleye), trout; *Marine*—bass (striped), clams, oysters, prawns, salmon (Atlantic), sea bass, worms; *Coordinator*—New York State Dept. of Agriculture & Markets, Field Operations Supervisor, Winners Circle, Albany, NY 12235.

NORTH CAROLINA: *Freshwater*—carp (koi), catfish, sunfish, ornamentals, tilapia; *Coordinator*—Div. of Aquaculture & Natural Resources, North Carolina Dept. of Agriculture, P.O. Box 27647, Raleigh, NC 27611.

NORTH DAKOTA: *Freshwater*—bait fish.

OHIO: *Freshwater*—bass (largemouth, white), carp, catfish, muskellunge, perch, pike (walleye), shad, suckers.

OKLAHOMA: *Freshwater*—bass (largemouth), catfish, tilapia; *Coordinator*—Oklahoma Dept. of Agriculture, Animal Industry Div., 2800 Lincoln, Oklahoma City, OK 73105.

OREGON: *Freshwater*—bass (largemouth), bluegill, crappie, crayfish, red drum, trout; *Marine*—clams, oysters, salmon; *Coordinator*—Oregon Dept. of Agriculture, Agriculture Development & Marketing Div., 121 SW Salmon St., Suite 240, Portland, OR 97204.

PENNSYLVANIA: *Freshwater*—bass (largemouth), bluegills, catfish, carp, crayfish, frogs, goldfish, pike (northern), hellgrammites (aquatic dobsonfly larvae, used as bait), leeches, muskellunge, trout; *Coordinator*—Bureau of Agricultural Development, 2301 N. Cameron St., Harrisburg, PA 17110.

PUERTO RICO: *Freshwater*—giant river prawn, tilapia; *Marine*—ornamentals, queen conch, seaweeds, shrimp (*Penaeus*).

RHODE ISLAND: *Marine*—clams; *Coordinator*—Div. of Fish & Wildlife, Coastal Fisheries Laboratory, 1231 Succotash Rd., RR 1, Wakefield, RI 02879.

SOUTH CAROLINA: *Freshwater*—bass (hybrids, largemouth, smallmouth, striped), bluegills, carp, catfish, crayfish, giant river prawn, redbreast (sunfish sp.), shad, shellcracker (redear sunfish), terrapin, trout; *Marine*—hard clams, oysters, red drum, shrimp (*Penaeus*), spotted seatrout; *Coordinator*—Marketing Specialist, P.O. Box 11280, Columbia, SC 29211.

SOUTH DAKOTA: *Freshwater*—bass (largemouth), paddlefish, trout.

TENNESSEE: *Freshwater*—bass (largemouth, striped), bluegills, carp (grass), catfish, minnows, pearl mussels, trout, turtles; *Coordinator*—Tennessee Dept. of Agriculture, Assistant Director of Marketing, Box 40627, Melrose Station, Nashville, TN 37204.

TEXAS: *Freshwater*—aquatic plants, bait fish, bass (largemouth), catfish, crappie, crayfish, pearl mussels, striped bass, sunfish, tilapia; *Marine*—brine shrimp, prawns, shrimp (*Penaeus*), striped bass (hybrid), oysters, red drum, spotted seatrout; *Coordinator*—Texas Dept. of Agriculture, Marketing Specialist, P.O. Box 12847, Austin, TX 78711.

UTAH: *Freshwater*—carp, trout, turtles; *Marine*—brine shrimp (*Artemia*, hypersaline); *Coordinator*—Utah Dept. of Agriculture, Director of Animal Industry & State Veterinarian, 350 North Redwood Rd., Salt Lake City, UT 84116.

VERMONT: *Freshwater*—trout; *Coordinator*—Vermont Dept. of Agriculture, Food & Markets, Agricultural Marketing Representative, 116 State St., Montpelier, VT 05602.

VIRGINIA: *Freshwater*—bass, bluegills, catfish, trout; *Marine*—bass (striped), clams, oysters; *Coordinator*—Virginia Dept. of Agriculture & Consumer Services, Div. of Marketing, P.O. Box 1163, Richmond, VA 23209.

VIRGIN ISLANDS (U.S.): *Marine*—seaweeds, tilapia.

WASHINGTON: *Freshwater*—carp, tilapia; *Marine*—abalone, algae, bass (striped), clams, mussels, oysters, prawn (spot), salmon, trout; *Coordinator*—Washington State Dept. of Agriculture, Aquatic Farm Program Manager, 406 General Administration Bldg., KU-14, Olympia, WA 98504.

WEST VIRGINIA: *Freshwater*—catfish, trout; *Coordinator*—West Virginia Dept. of Agriculture, Director, Marketing & Development, State Capitol, Charleston, WV 25305.

WISCONSIN: *Freshwater*—bass (largemouth), muskellunge, perch (yellow), pike (walleye), trout.

WYOMING: *Freshwater*—muskellunge, trout; *Coordinator*—Wyoming Dept. of Agriculture, Marketing Director, 2219 Carey Ave., Cheyenne, WY 82002.

NEIGHBORING NORTH AMERICAN REGIONS

BAHAMAS: *Freshwater*—ornamentals (clownfish, neon goby); *Marine*—bass (striped), Florida pompano, ornamentals (angel fish, butterfly fish, clownfish, etc.), pinfish, red drum, shrimp (*Penaeus*), tilapia.

BERMUDA: *Freshwater*—carp (grass); *Marine*—dolphin fish, queen conch.

BONAIRE (NETHERLANDS ANTILLES): *Freshwater*—ornamentals; *Marine*—giant clams, shrimp (*Penaeus*), spiny lobster, tilapia.

CANADA: *Freshwater*—muskellunge, perch (yellow); *Marine*—abalone, algae (Irish moss), char (Arctic), lobster (American), mussels, oysters, salmon (Atlantic, Pacific), scallops, trout.

CENTRAL AMERICA: *Freshwater*—catfish, giant river prawns, trout; *Marine*—shrimp (*Penaeus*).

CUBA: *Freshwater*—carp, catfish, tilapia; *Marine*—groupers, mullet, oysters (mangrove), shrimp, snappers.

DOMINICAN REPUBLIC: *Freshwater*—giant river prawn, tilapia; *Marine*—Florida pompano, shrimp (*Penaeus*).

GREAT LAKES: char (Arctic), perch (yellow), pike (walleye), salmon.

MARTINIQUE: *Marine*—permit, red drum, sea bass, sea bream, shrimp (*Penaeus*), snappers, tilapia.

MEXICO: *Freshwater*—carp, giant river prawn, trout; *Marine*—clams, mullet, oysters, pen shell, scallops, shrimp (*Penaeus*).

TURKS AND CAICOS: *Marine*—queen conch.

Appendix F

Examples of Important Predators on Aquacultural Species

A starfish opens an oyster to devour the meat.

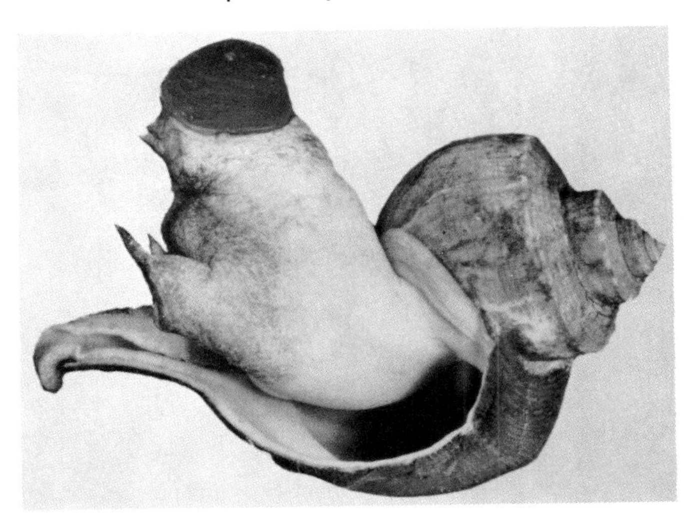

Once a whelk's prey (usually a bivalve) opens its shell, the whelk inserts the edge of its own shell into the prey. This is followed by the whelk's *radula* (armed proboscis), which helps in consuming the soft flesh.

An oyster drill bores through the shell of a young oyster, then eats the meat. Drills also eat mussels and barnacles.

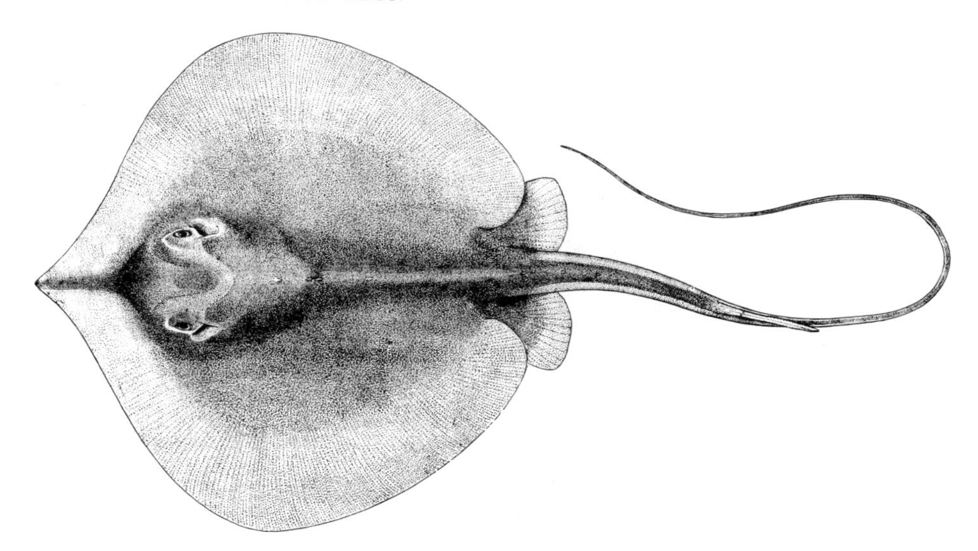

Rays eat bottom-dwelling mollusks, crabs, and fishes.

The heron (*facing page*) is just one of the many birds that prey on aquacultural species: Gulls, terns, mergansers, kingfishers, cormorants, ospreys, pelicans, and ducks can also be serious predators under certain circumstances. The increase in water areas under cultivation has increased the food supply for fish-eating birds, and is believed to have augmented their numbers.

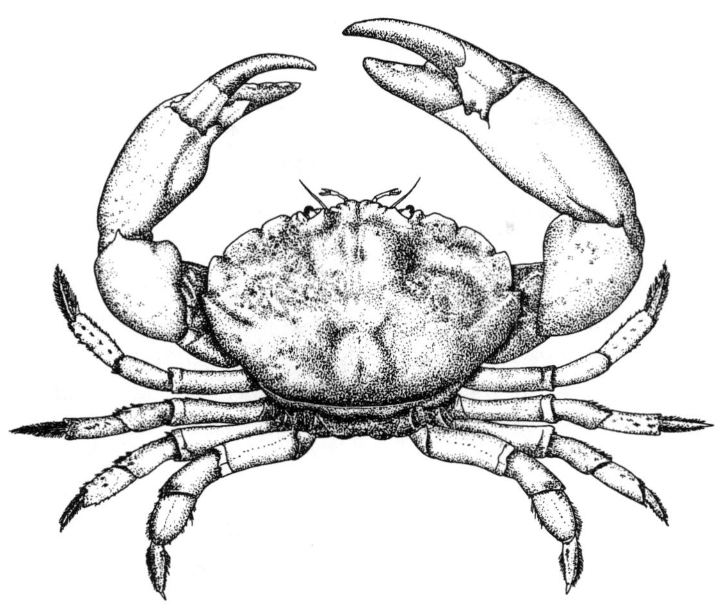

Stone crabs eat polychaetes, bivalves, mollusks (e.g., oysters, clams, and conch), and crustaceans (e.g., blue crabs, hermit crabs); they are also cannibalistic.

Appendix G

Examples of Important Parasites of Aquacultural Species

Parasites of aquacultural species infect or infest nearly every organ or surface of fishes (*see facing page*). Their life cycles are either *direct,* from one fish to another (as for "Ich," *below*), or *indirect,* with one or more intermediate hosts (as for digenetic trematodes, p. 295).

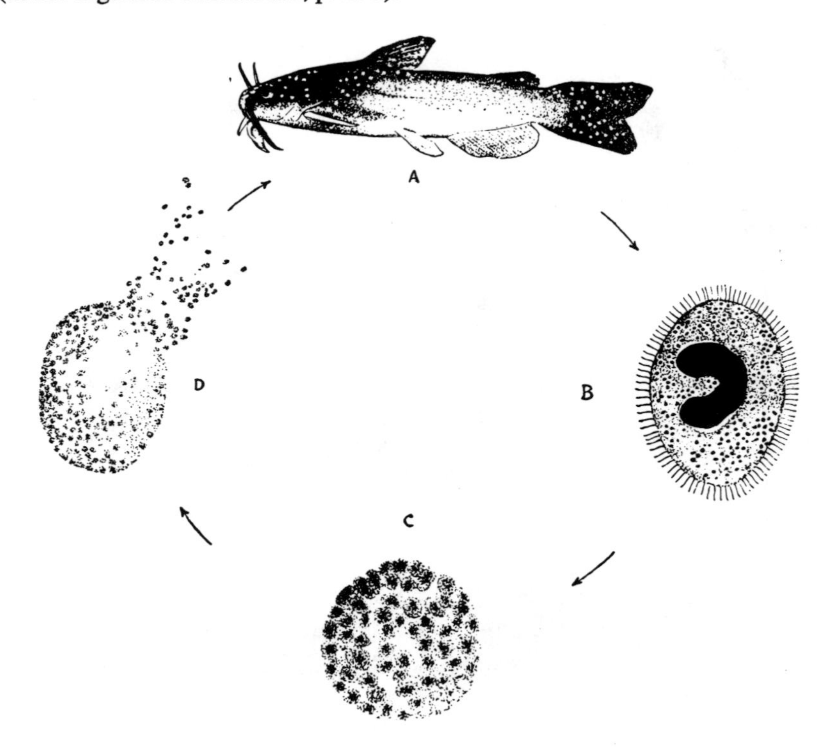

Life cycle of *Ichthyophthirius* ("Ich" for short), an ectoparasitic protozoan on warm-water fishes (especially catfish), trout, and aquarium fishes: (A) White spots and swellings in the epidermis of a catfish are caused by the parasite; (B) free-swimming stage leaves the host and settles to the bottom; (C) parasite divides into many small individuals in cyst; (D) cyst ruptures, releasing hundreds of minute parasites to infect or reinfect fish hosts.

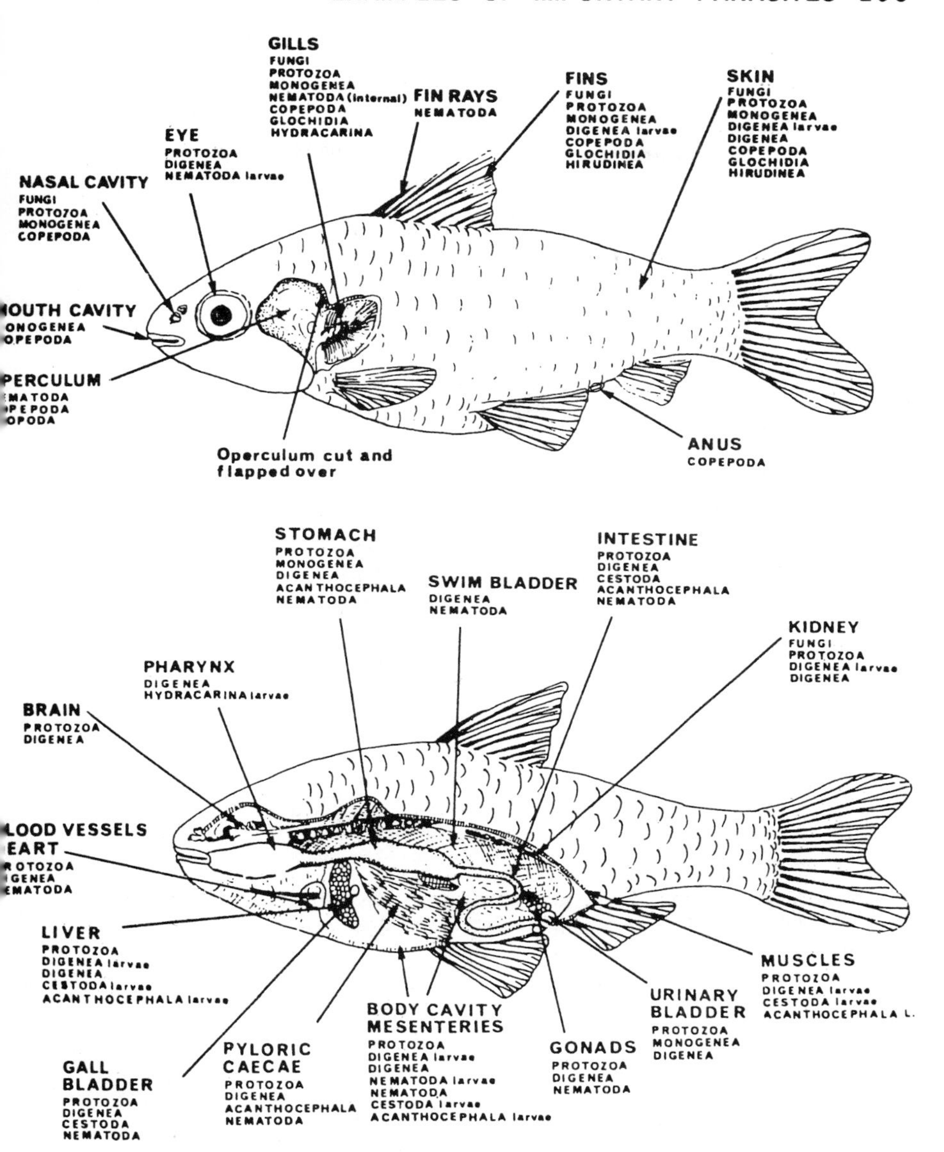

Diagrams of fishes showing sites of infestation and infection.

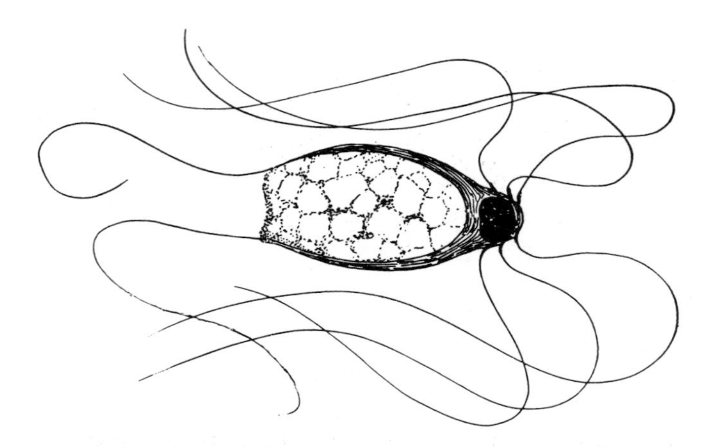

Hexamita, a protozoan with eight flagella that is found in oysters and in the intestines of fish. Although rather common, its pathogenicity is not well documented.

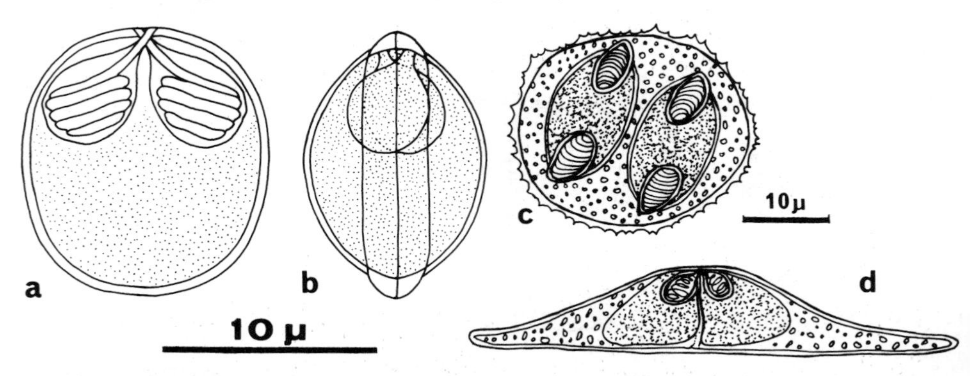

Myxosporeans, spore-forming endoparasitic protozoans that infect fish: (a) front view of spore showing coiled polar filaments; (b) side (*sutural*) view of spore; (c) growing stage (*trophozoite*) containing two developing spores; (d) spore showing the *sporoplasm*—the infective portion of the spore that leaves the spore case to infect another fish—near the polar filament.

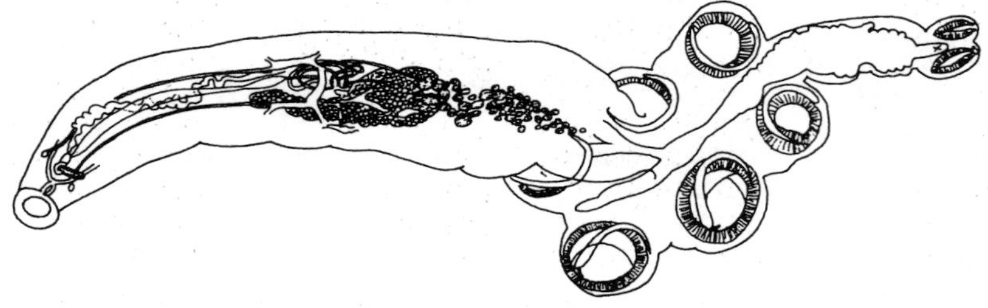

Monogenetic trematode, an ectoparasite with a direct life cycle. It is found on the skin and gills of fish, interfering with respiration and osmotic balance.

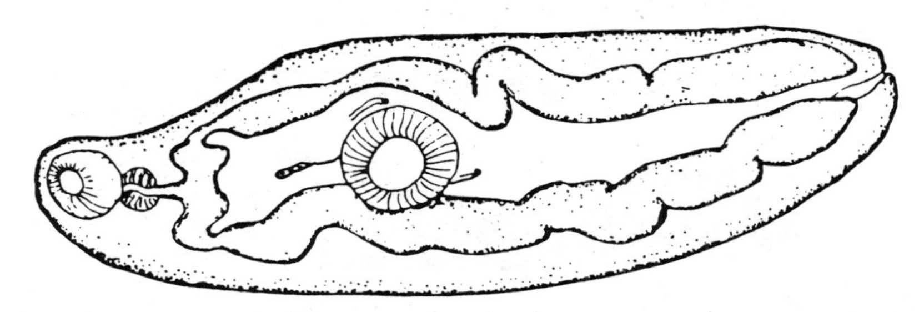

Digenetic trematode, a parasitic flatworm with suckers and hooks. Its life cycle may require several hosts: Adults (shown) usually occupy hollow organs of fish and some shellfish; larvae invade the flesh and organs of fishes and shellfish.

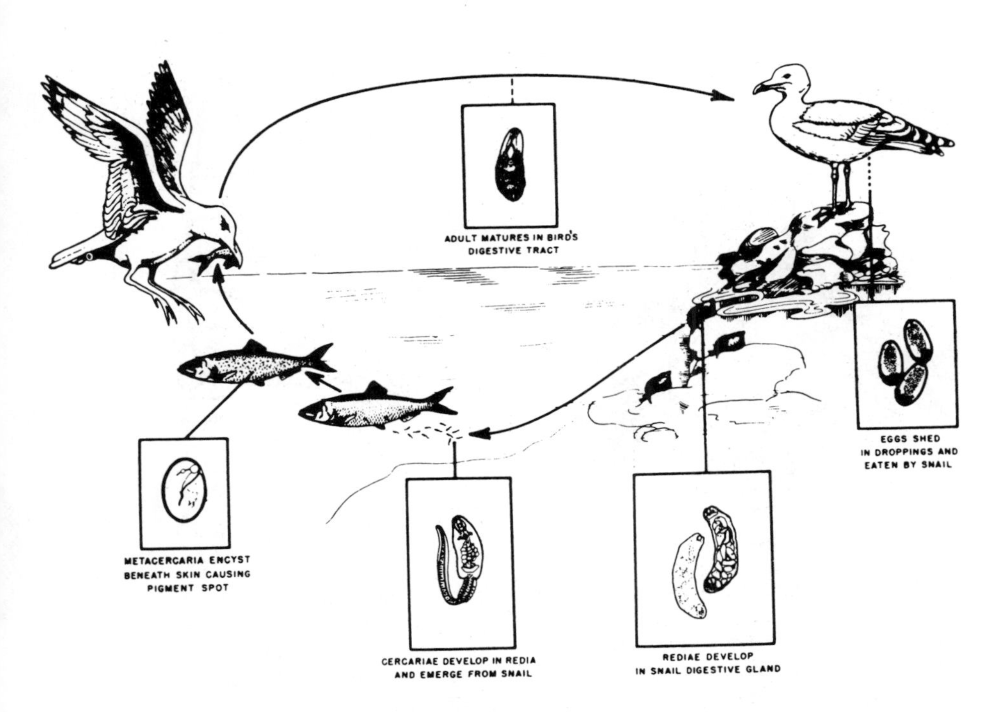

Life cycle of a digenetic trematode, showing the several intermediate hosts through which these parasites pass in order to mature.

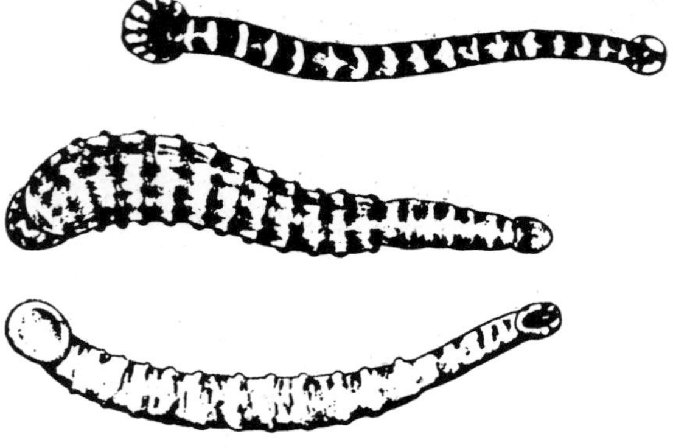

Leeches, parasitic or predatory annelid worms that may serve as a vector transporting blood parasites.

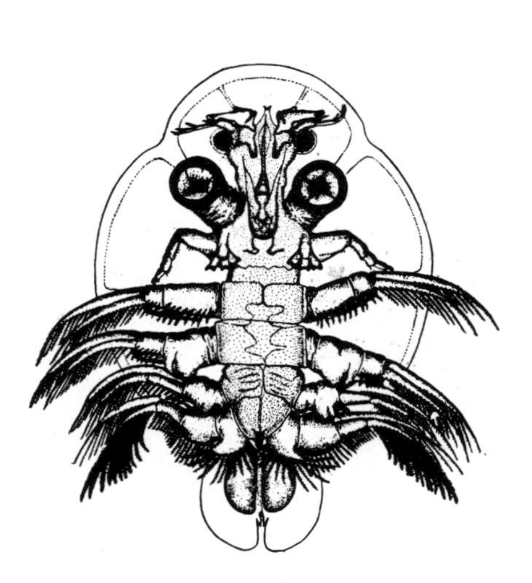

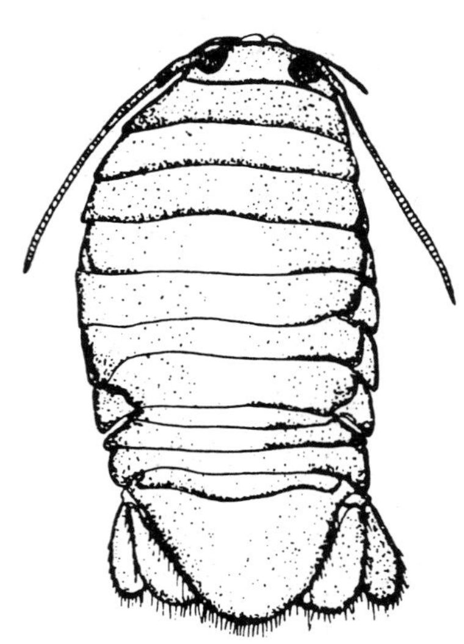

Fish lice (Branchiurans) are crustaceans represented by a number of genera, including *Argulus.* They skitter about on the skin of fish (mostly freshwater species), causing them to scrape on the bottom, or on sharp objects in tanks or ponds, and thus become infected by bacteria and fungi.

Isopods, small marine or freshwater ectoparasitic crustaceans, attach to the tongues or heads of some fishes by means of sharp claws, then share in their hosts' food.

Appendix H

National Aquaculture Associations

American Alligator Farmers Association
5145 Harvey Tew Rd.
Plant City, FL 33565
813-754-3595

American Fish Farmers Federation
P.O. Box 161
Lonoke, AR 72086
501-676-2800

American Fisheries Society
5410 Grosvenor Lane, Suite 110
Bethesda, MD 20814
301-897-8616

American Tilapia Association
Kirkwood Rural Development Center
P.O. Box 2068
Cedar Rapids, IA 52406
319-398-5699

Aquaculture Association of Canada
P.O. Box 1987
St. Andrews, NB E0G 2X0
Canada
506-529-8854

Associated Koi Clubs of America
340 Mariposa Drive
Camarillo, CA 93010
805-482-0556

Caribbean Aquaculture Association
Dept. of Marine Sciences
University of Puerto Rico
Mayaguez, PR 00708

Catfish Farmers of America
1100 Highway 82 East, Suite 202
Indianola, MS 38751
601-887-2699

International Association of Astacology
P.O. Box 44650 (Univ. of Southwestern Louisiana)
Lafayette, LA 70504
318-231-5239

National Aquaculture Association
P.O. Box 220
Harpers Ferry, WV 25425
304-876-6666

National Fisheries Institute Aquaculture Council
1525 Wilson Blvd., Suite 500
Arlington, VA 22209
703-524-8884

National Ornamental Goldfish Growers Association
6916 Black's Mill Road
Thurmont, MD 21788
301-271-7475

National Shellfisheries Association
Univ. New Orleans, Dept. of Biological Sciences
New Orleans, LA 70148
504-286-7042

Shellfish Farmers Association
480 River Prado Road
Ft. Pierce, FL 34946
407-466-2013

Shellfish Institute of North America
1525 Wilson Blvd., Suite 500
Arlington, VA 22209
202-524-8883

Striped & Hybrid Bass Producers' Association
P.O. Box 8605
Raleigh, NC 27695
919-737-2454

United States Trout Farmers
P.O. Box 220
Harpers Ferry, WV 25425
304-876-6666

World Aquaculture Society
143 John M. Parker Coliseum
Louisiana State University
Baton Rouge, LA 70803
504-388-3137

Appendix I

Measurement Conversion Table

	Multiply	By	To Obtain
Metric to U.S. Customary	millimeters (mm)	0.03937	inches
	centimeters (cm)	0.3937	inches
	meters (m)	3.281	feet
	meters (m)	0.5468	fathoms
	kilometers (km)	0.6214	statute miles
	kilometers (km)	0.5396	nautical miles
	square meters (m^2)	10.76	square feet
	square kilometers (km^2)	0.3861	square miles
	hectares (ha)	2.471	acres
	liters (l)	0.2642	gallons
	cubic meters (m^3)	35.31	cubic feet
	cubic meters (m^3)	0.0008110	acre-feet
	milligrams (mg)	0.00003527	ounces
	grams (g)	0.03527	ounces
	kilograms (kg)	2.205	pounds
	metric tons (t)	2205.0	pounds
	metric tons (t)	1.102	short tons
	kilocalories (kcal)	3.968	British thermal units
	Celsius degrees (^oC)	$1.8(^oC) + 32$	Fahrenheit degrees
U.S. Customary to Metric	inches	25.40	millimeters
	inches	2.54	centimeters
	feet (ft)	0.3048	meters
	fathoms	1.829	meters
	statute miles (mi)	1.609	kilometers
	nautical miles (nmi)	1.852	kilometers
	square feet (ft^2)	0.0929	square meters
	square miles (mi^2)	2.590	square kilometers
	acres	0.4047	hectares
	gallons (gal)	3.785	liters
	cubic feet (ft^3)	0.02831	cubic meters
	acre-feet	1233.0	cubic meters
	ounces (oz)	28350.0	milligrams
	ounces (oz)	28.35	grams
	pounds (lb)	0.4536	kilograms
	pounds (lb)	0.00045	metric tons
	short tons (ton)	0.9072	metric tons
	British thermal units (Btu)	0.2520	kilocalories
	Fahrenheit degrees (^oF)	$0.5556\ (^oF - 32)$	Celsius degrees

Subject Index

Common names of aquacultural species are most often indexed by keyword noun (e.g., largemouth *bass, buffalo* fish, Arctic *char,* red *drum*); but names spelled as one word (e.g., bluegill) are alphabetized as such. Italic page numbers indicate illustrations.

Taxonomic Index